The Digital Practice

Campion Quinn

The Digital Practice

Proven Marketing Tactics for Today's Physician

Springer

Campion Quinn
Rockville Medical, LLC
Rockville Centre, NY, USA

ISBN 978-3-032-07744-8 ISBN 978-3-032-07745-5 (eBook)
https://doi.org/10.1007/978-3-032-07745-5

This Springer imprint is published by the registered company Springer Nature Switzerland AG
The registered company address is: Gewerbestrasse 11, 6330 Cham, Switzerland

For my beloved children, Paige, Campion, Jr., and Alannah—the greatest gifts I've ever received.

Contents

Chapter 1
Introduction to Digital Marketing for Healthcare Providers

The Shift to Digital-First Healthcare

Technology has changed how patients find and evaluate care. According to the CDC's National Health Interview Survey, 58.5% of US adults used the Internet to search for health or medical information in 2022 [1]. A 2023 Healthgrades consumer survey found that 72% of patients begin their search for a provider online, and 60% report that online reviews influence their decision-making process [2]. Finally, a 2024 Pew Research Center study found that 83% of adults now search online before choosing a healthcare provider [3]. These trends have made digital visibility a strategic priority for every healthcare practice.

These findings are illustrated in Fig. 1.1, which shows recent trends in online health information seeking among US adults.

A strong online presence enables providers to clearly describe their services, enhance patient access, and differentiate themselves in a crowded marketplace. Digital tools facilitate targeted communication, allowing providers to effectively address specific patient needs. These tools help build trust, improve patient satisfaction, and encourage patients to return.

Patients often begin their healthcare search by using Google, with studies showing many turn to online queries before consulting healthcare providers or others. Research shows that health-related searches on Google tend to increase significantly just before patients visit emergency departments, indicating that patients seek information about symptoms, treatments, and hospital logistics online prior to care.[1] Search engine algorithms determine which providers appear at the top of results.

[1] Penn Medicine. https://chti.upenn.edu/utilizing-google-search-data-gain-insight-health. Accessed 8/8/25.

C. Quinn, *The Digital Practice*, https://doi.org/10.1007/978-3-032-07745-5_1

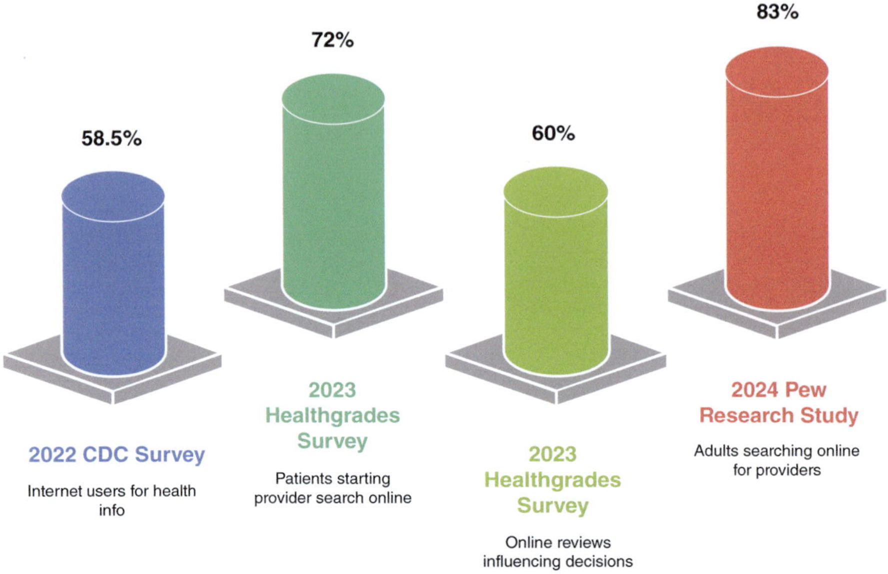

Fig. 1.1 Online health information seeking trends (Quinn/Napkin.AI)

Those with complete and verified profiles,[2] recent reviews,[3] mobile-optimized websites,[4] and locally relevant content rank higher.[5] If a practice does not appear on the first page of search results, patients are less likely to choose it. Search visibility serves as a core element of patient acquisition.

Online reputation plays an equally important role. A 2023 Healthgrades survey reported that 90% of patients read online reviews before selecting a provider. Most patients will not consider providers with an average rating below 4.0 [2]. The number of reviews and their recency also influence patient choices. Public feedback on platforms such as Google, Healthgrades, and Yelp helps shape how prospective patients perceive clinical quality.

As a result, reputation management is a crucial component of any effective digital strategy. Providers must monitor review platforms, respond to patient feedback, and encourage satisfied patients to share their experiences. A steady stream of recent, authentic reviews can help establish credibility and attract new patients [4].

Digital engagement is no longer optional. It is now essential to build relationships, increase visibility, and maintain relevance in an online-first healthcare

[2] https://mhp.si/seo-best-practices-for-healthcare-organizations/. Accessed 8/7/25.

[3] https://growthnatives.com/blogs/digital-marketing/the-vital-pulse-of-health-care-seo-how-search-visibility-drives-practice-success/. Accessed 8/7/25.

[4] https://www.wheelhousedmg.com/insights/articles/the-ultimate-guide-to-healthcare-seo/. Accessed 8/7/25.

[5] Cardinal. https://www.cardinaldigitalmarketing.com/healthcare-resources/blog/local-seo-ranking-factors-healthcare/. Accessed 8/7/25.

system. Later chapters will explain how to use specific digital tools and techniques to help healthcare providers adapt to these changes and support stronger patient connections. To stay competitive, your practice must appear in the same digital spaces where patients search for care, starting with a well-structured website and a visible presence on Google.

Why Every Healthcare Practice Needs a Digital Strategy

As mentioned above, patients now turn to the Internet as their primary source of health information. This shift prompts healthcare providers to adopt digital marketing strategies to meet patients where they are: online. Digital marketing allows providers to connect with potential patients, build trust, and establish a robust online presence.[6]

Digital marketing ensures that physicians are present and accessible where patients are searching for them. It plays a pivotal role in positively influencing patients' decisions by providing them with the information they need to make informed choices about their healthcare.[7] Through strategies such as healthcare social media marketing, physicians can participate in community discussions, network, and promote important health information, thereby playing a crucial role in public health education.[8]

In addition to visibility, digital marketing now plays a direct role in shaping patient experience by employing tools that directly improve it. These "digital front door" tools, such as online scheduling, automated follow-up, and real-time reputation management, reduce friction and create a smoother patient journey [5]. Platforms like NexHealth [6] and Doctible [7] enable patients to book appointments, receive reminders, and leave feedback without needing to call the office. When used in conjunction with HIPAA-compliant CRMs, these tools enhance trust, boost retention, and facilitate efficient practice operations [8].

Moreover, digital marketing tools like video marketing can help the healthcare industry become more approachable, credible, and reliable. By sharing informative and engaging content, such as animated explainer videos, patient story reels, and interactive quizzes, physicians can demystify complex medical topics, making it easier for patients to understand their health conditions and the treatment options available to them.[9] An effective digital marketing strategy involves a deep

---

[6] https://www.ednadigitalmarketing.com/why-digital-marketing-matters-for-medical-practice. Accessed 3/17/24.

[7] https://www.linkedin.com/pulse/digital-marketing-doctors-prescription-healthcare-success-al-dic/. Accessed 3/17/24.

[8] https://webmdignite.com/blog/evolving-role-digital-marketing-healthcare. Accessed 3/17/24.

[9] https://www.seattlenewmedia.com/blog/digital-marketing-for-healthcare-industry. Accessed 3/17/24.

understanding of patients' needs, identifying latent needs, and offering new health solutions that resonate with the target audience [9].

Digital platforms help promote health information, but they also carry risks. Misinformation can spread rapidly, particularly in areas such as vaccines, treatments, and emerging diseases. Healthcare providers must ensure that their messaging is accurate, up-to-date, and evidence-based. By doing so, they serve as reliable voices in a noisy digital environment, helping patients navigate complex health topics safely and effectively.

The healthcare marketer's role in the digital patient journey is to provide all the necessary information patients need to find and choose their healthcare provider. This involves creating content that addresses patients' concerns at every stage of their healthcare journey, from symptom research to posttreatment care.[10] By doing so, digital marketing not only enhances a practice's reputation but also significantly expands its patient base.

Digital marketing is not just an option for physicians; it is an imperative. It is the key to building a practice's reputation and expanding its patient base in a world where patients increasingly rely on digital channels for healthcare information and services. Start by auditing your current online footprint. Search for your practice as a patient would and note what's missing, outdated, or hard to find.

A successful strategy depends not only on visibility but also on how data is collected and utilized to personalize patient communication while respecting patient privacy.

Sidebar: Top 5 Reasons Patients Choose Providers Online

1. **Positive Reviews and Ratings**
   Patients look for recent, detailed feedback that signals trust, professionalism, and quality care.
2. **Clear, Professional Website**
   A well-designed site with bios, photos, services, insurance info, and online booking increases confidence.
3. **Online Scheduling Options**
   Easy-to-use appointment tools are a major deciding factor, especially for busy or tech-savvy patients.
4. **Strong Local Presence**
   Providers who appear in Google's local 3-pack or map listings gain more clicks and visits.
5. **Helpful, Trustworthy Content**
   Blog posts, videos, and FAQs that answer real patient questions help establish authority and build rapport before the first visit.

[10] https://www.yext.com/blog/2024/01/what-is-marketings-role-in-the-digital-patient-journey. Accessed 3/17/24.

Understanding Zero-Party Data in Healthcare Marketing

Zero-party data is information that patients intentionally and proactively share with a healthcare provider. This may include responses to online health quizzes, symptom checkers, intake forms, preference settings, or interactions with chatbots. Unlike first-party data, which is passively collected through behaviors like page views or clicks, zero-party data reflects the patient's explicit choices and interests. It is voluntarily offered, which makes it more accurate, relevant, and compliant with privacy regulations.

For healthcare marketers, zero-party data is especially valuable. It enables personalized communication while respecting consent and transparency, both of which are essential in regulated environments like HIPAA and CCPA. For example, a patient who completes a skin health quiz may receive targeted content on dermatology services. A visitor who selects "email only" in their communication preferences will not receive texts. These personalized experiences foster trust, minimize unwanted outreach, and enhance campaign performance. Zero-party data also provides a practical alternative to third-party tracking, which is becoming less reliable due to browser restrictions and privacy laws. Table 1.1 summarizes the different types of patient data used in digital marketing, along with examples and associated privacy considerations.

Callout Box: Types of Patient Data in Digital Marketing

Type of data	Definition	Examples	Privacy level
Zero-party data	Information the patient *proactively and voluntarily* shares	Quiz answers, symptom checkers, preference settings, surveys	Highest (consensual and explicit)
First-party data	Information collected *passively* through patient behavior on your platforms	Website visits, appointment bookings, email opens	High (collected by your site or system)
Third-party data	Information collected *by outside companies* and shared or sold to others	Purchased demographic lists, ad network tracking cookies	Lowest (privacy concerns, now declining in use)

Tip: Use zero- and first-party data to personalize outreach while staying HIPAA- and CCPA-compliant. Avoid relying on third-party data due to the growing legal and ethical risks associated with it. Use tools like surveys, quizzes, or chatbot prompts to invite patients to share preferences and interests. Keep consent clear, and tailor your messaging accordingly.

Fig. 1.2 The various marketing channels that are used in healthcare (Quinn/ChatGPT)

Once a practice understands what types of data to collect, the next step is choosing the right channels to deliver tailored messages and services.

Figure 1.2 shows the variety of marketing channels available for medical practices.

Top Digital Channels for Reaching Patients

Understanding the landscape of digital marketing channels and their application to healthcare is essential for healthcare providers seeking to effectively reach and engage their patient base. A channel refers to the medium or platform through which a company communicates or delivers its marketing messages, products, or services to its target audience. These channels enable businesses to reach potential customers in the digital space, encompassing a variety of platforms, including search engines, social media, email, websites, and other online platforms. Each channel serves a specific purpose and can be utilized in different ways to engage with audiences, distribute content, promote products or services, and ultimately drive conversions and achieve marketing objectives.[11,12]

Physicians can utilize multiple digital channels to connect with patients, from raising awareness to scheduling appointments. Each medium offers a distinct approach to informing, reassuring, and engaging patients.

Search Engine Marketing encompasses both search engine optimization (SEO) and paid advertising (PPC). SEO makes your website easier to find by matching content to patient queries. Paid ads, such as Google Ads, place your practice at the top of search results.[13] PPC, on the other hand, allows advertisers to bid on keywords, ensuring their ads appear in search engine results and pay a fee each time the ad is clicked.[14] For example, Google Ads is a popular platform for PPC campaigns.

Social Media Marketing reaches audiences on Facebook, Instagram, LinkedIn, X, and TikTok. One survey found that 56% of Gen Z use TikTok for health advice, and one-third cite it as their primary source of wellness information.[15] Brief educational videos or patient stories on TikTok and Instagram can help build credibility and foster connections.

Email Marketing helps retain patients by delivering appointment reminders, follow-up tips, or health newsletters. Segmented emails enable more relevant and personalized communication. Tools like Mailchimp enable businesses to segment their audience and automate campaigns for efficiency.

Display Advertising uses banner or video ads on websites that visitors frequent. These ads can effectively promote seasonal services such as flu vaccines or highlight specialties, and they function well when combined with search and email campaigns.

Content Marketing earns patient trust through useful materials such as blog posts, explainer videos, podcasts, and infographics. When content answers patient questions, it establishes your authority before they ever step into the office.

Influencer Marketing can involve collaboration with trusted health personalities, such as physician content creators, patient advocates, or wellness coaches. These partnerships can lend authenticity to your message.

---

[11] https://mailchimp.com/resources/what-are-marketing-channels/. Accessed 3/18/24.

[12] https://www.semrush.com/blog/digital-marketing-channels/. Accessed 3/18/24.

[13] https://searchengineland.com/guide/what-is-seo. Accessed 3/18/24.

[14] https://searchengineland.com/guide/what-is-paid-search. Accessed 3/18/24.

[15] https://www.healthline.com/health-news/tiktok-main-source-health-information-gen-z?utm_source=chatgpt.com. Accessed 7/23/25.

Mobile marketing Is Essential With roughly 64% of site visits coming from mobile devices, optimizing your site for phones and tablets, including SMS reminders and mobile-friendly pages, improves access and engagement.[16,17]

Emerging Tools

AI-Powered SEO now includes optimization for Google's Search Generative Experience (SGE), which displays AI-generated summaries alongside search results.[18] Medical practices can improve their visibility by utilizing clear, structured content formats, such as FAQs, headings, and schema markup. Schema markup is a type of code that labels each part of a webpage (such as addresses, phone numbers, or services) so that search engines can better understand and display that information.

Programmatic Advertising uses automated, real-time bidding to place ads in front of the right patient, across device types, and at scale.[19]

Voice Search Optimization adapts content to how patients speak to Siri, Alexa, or Google Assistant. Studies show 32% of patients use voice search to find providers.[20] Listing accurate address and phone details helps capture voice-driven traffic.

These channels work best as parts of a coordinated strategy. In the section below, we'll show how to integrate them into a multichannel plan that aligns with your practice goals and patient habits. Select two to three digital channels based on your patient demographics. Focus your efforts there before expanding to additional platforms.

Table 1.2 ranks preferred digital channels by patient age group, highlighting how different demographics engage online.

Channel Ranking By Age Range

Rank	Ages 18–27 (Gen Z)	Ages 28–43 (Millennials)	Ages 44–59 (Gen X)	Ages 60+ (Boomers)
1	TikTok	Facebook	Facebook	Email
2	Instagram	Instagram	SEO	SEO

(continued)

[16] https://wgcontent.com/blog/healthcare-marketing-and-tiktok/?utm_source=chatgpt.com

[17] https://gs.statcounter.com/platform-market-share/desktop-mobile/united-states-of-america

[18] https://kanopi.com/blog/googles-search-generative-experience-101/?utm_source=chatgpt.com

[19] https://www.insivia.com/programmatic-advertising-statistics/?utm_source=chatgpt.com

[20] https://www.synup.com/en/voice-search-statistics?utm_source=chatgpt.com

Rank	Ages 18–27 (Gen Z)	Ages 28–43 (Millennials)	Ages 44–59 (Gen X)	Ages 60+ (Boomers)
3	YouTube	YouTube	Email	Facebook
4	Mobile marketing (SMS)	Email	Paid search (Google ads)	YouTube
5	SEO	SEO	Voice search optimization	Online reviews
6	Voice search optimization	Voice search optimization	Online reviews	Patient portals and CRM
7	Online reviews	Paid search	Patient portals and CRM	Display advertising
8	Email	Content marketing	Content marketing	Content marketing
9	Display advertising	Online reviews	Display advertising	Voice search optimization

Key Patterns

- **Younger users (18–27)** prioritize **visual and short-form video content** and spend more time on social platforms like TikTok and Instagram.
- **Millennials (28–43)** use **social and email** actively and are very responsive to mobile communication and targeted ads.
- **Gen X (44–59)** prefers **search-based** tools, **email**, and reliable content sources.
- **Boomers (60+)** respond best to **direct, clear communication** like email, trusted search results, and reviews.

Building an Inclusive Multichannel Approach

Multichannel Marketing refers to the coordinated use of multiple digital platforms to reach patients at various touchpoints, such as websites, search engines, social media, and email. Rather than relying on one primary channel, healthcare providers use a combination to deliver consistent messages, promote services, and support patient engagement across devices and preferences.[21]

Each channel plays a distinct role:

- **Search engines** (SEO/PPC): Help patients find the practice during active research.
- **Social media**: Encourages real-time interaction and community-building.
- **Email**: Supports follow-up, appointment reminders, and patient education.
- **Display ads**: Promote seasonal campaigns like flu shots or screenings.
- **Mobile tools**: Ensure accessibility across smartphones and tablets.
- **Content marketing**: Builds trust by answering patient questions before a visit.

[21] https://www.techtarget.com/searchcustomerexperience/definition/multichannel-marketing. Accessed 3/18/24.

A multichannel approach works best when these platforms reinforce one another. For example, a Facebook ad may direct users to a landing page, where a patient fills out a contact form and later receives a follow-up email.

The goal is to create a cohesive and integrated marketing strategy that reaches consumers across various touchpoints, providing them with a seamless experience regardless of the channel or device they use. By employing multichannel marketing, healthcare organizations aim to expand their reach, enhance patient engagement, and ultimately achieve better health outcomes and business success.[22,23] These coordinated efforts expand reach, improve personalization, and support a seamless patient experience.

A multichannel strategy enables healthcare providers to reach a broader range of patients across various platforms. It allows them to cater to diverse patient demographics, addressing their unique preferences and needs. For instance, younger patients may be more accessible through social media platforms, while older demographics might respond better to email marketing or traditional search engine optimization (SEO) tactics. By employing a multichannel strategy, healthcare providers can ensure that their messages are delivered through the most appropriate and effective channels for each patient group.[24]

Moreover, a multichannel approach facilitates inclusivity, ensuring that healthcare marketing efforts reach a broad spectrum of the patient population, including those from various cultural and linguistic backgrounds. Inclusivity is crucial in a healthcare setting, where understanding and addressing the diverse needs of patients can significantly impact health outcomes. Digital marketing strategies that emphasize inclusivity and diversity can help healthcare practices build trust and establish strong relationships with their patients, ultimately leading to better healthcare experiences for all [10].

Navigating the digital landscape requires healthcare providers to understand the key digital marketing channels and their relevance to the healthcare industry. Adopting a multichannel approach is vital for reaching diverse patient demographics, ensuring inclusivity, and effectively engaging with patients across the digital spectrum. Feature physician-created content regularly. Even short posts or quick videos answering common questions build authority and reassure patients.

Reaching patients is only part of the equation. To earn their trust, providers must share credible, relatable content rooted in medical expertise.

[22] https://www.practicebuilders.com/blog/multichannel-medical-marketing-highway-to-a-successful-practice/. Accessed 3/18/24.

[23] https://www.kiwihealth.com/blog/multi-channel-marketing-for-healthcare. Accessed 3/18/24.

[24] https://www.mdconnectinc.com/medical-marketing-insights/medical-marketing-insights/digital-marketing-diversity. Accessed 3/17/24.

Using Medical Expertise to Build Trust Online

In the realm of digital marketing, healthcare providers possess a unique advantage: their medical expertise. This expertise allows them to create authoritative content tailored to patients' needs. Examples include animated explainer videos that clarify complex procedures, interactive quizzes for symptom awareness, or short video reels featuring patient stories or staff introductions. By focusing on transparency, honesty, and the dissemination of valuable health information, healthcare providers can go beyond mere patient acquisition to build trust and credibility with their audience.[25]

Trust and credibility are the cornerstones of a successful online healthcare practice. One effective way to build this trust is using patient testimonials. By incorporating transparent testimonials into their digital marketing strategies, medical institutions can cultivate an environment of trust and reliability.

Testimonials provide authentic and compelling accounts of individuals' experiences with healthcare services, offering prospective patients insights into the quality of care they can expect.[26,27] However, providers must ensure these testimonials comply with HIPAA and FTC regulations. They should never include identifiable patient information without written consent, and any incentives offered for reviews must be clearly disclosed.

Moreover, showcasing professional achievements online further solidifies the trustworthiness and credibility of healthcare providers. This can include certifications, awards, or recognitions in their field, which reassure patients of the provider's competence and dedication to excellence in healthcare.

Leveraging medical expertise to create authoritative content, showcasing professional achievements, and utilizing patient testimonials are powerful strategies for healthcare providers in digital marketing. These approaches not only engage and educate patients but also build a foundation of trust and credibility essential for the growth and success of healthcare practices.

Healthcare providers increasingly use low-cost tools like user-generated content (UGC) and AI-assisted video creation to reach patients. UGC may include a patient describing their experience or a caregiver sharing a recovery story, posted with informed consent. Practices must follow HIPAA rules and FTC testimonial guidelines when using this material. AI tools, such as Synthesia or HeyGen, enable clinics to produce short, branded videos, such as how-to guides or wellness tips, without the need for a film crew or hiring a production team. These tools enable busy

---

[25] https://medium.com/@b27819792/healthcare-hero-building-trust-and-credibility-with-digital-marketing-0af7d0dcc7a3. Accessed 3/17/24.

[26] https://www.linkedin.com/pulse/patient-testimonials-tool-building-trust-healthcare-lamplugh-jr-rvoic/. Accessed 3/17/24.

[27] https://docstokes.com/blog/the-impact-of-patient-testimonials-in-healthcare-marketing. Accessed 3/17/24.

practices to publish consistent, high-quality content that supports patient education and fosters trust.

Trust also depends on how responsibly a practice handles patient data. Digital outreach must align with HIPAA and other privacy regulations.

HIPAA, Data Privacy, and Compliance in Marketing

In this new digital environment, healthcare providers face the dual challenge of navigating privacy concerns and regulatory compliance, particularly with laws such as the Health Insurance Portability and Accountability Act (HIPAA). HIPAA sets the standard for protecting sensitive patient data, requiring healthcare entities to implement physical, network, and process security measures. Understanding and adhering to these regulations is crucial for maintaining patient trust and avoiding legal repercussions.[28]

In addition to HIPAA, practices that collect patient data online must also comply with other privacy laws such as the California Consumer Privacy Act (CCPA).[29] CCPA applies to organizations that collect personal data from California residents. It requires businesses to inform users how their data is used, shared, and stored. This law affects healthcare marketing when practices use tracking tools or collect identifiable data from visitors in California.

Many physician websites utilize tracking technologies, such as Meta Pixel or Google Analytics, to monitor how visitors interact with their pages. These tools collect data on page views, clicks, and user behavior to help practices understand which services patients are researching, how they found the website, and whether they return. Marketers use this information to improve digital advertising, tailor content, and measure campaign effectiveness. When configured without proper safeguards, these trackers may also collect sensitive patient information.

The US Department of Health and Human Services (HHS) has warned that some common tracking technologies may violate HIPAA [11]. Meta Pixel and Google Analytics, when placed on patient portals, appointment pages, or symptom checkers, may collect and transmit protected health information (PHI) to third parties. A 2023 *STAT News* investigation found that dozens of hospitals had installed these tools on pages where patients entered symptoms or booked visits [12]. HHS guidance now recommends removing these trackers unless they meet strict privacy and consent requirements [13].

To maintain both compliance and efficiency, many practices now use HIPAA-compliant customer relationship management (CRM) tools. Platforms such as Salesforce Health Cloud and Klara allow physicians to send appointment

[28] https://www.hhs.gov/hipaa/for-professionals/security/laws-regulations/index.html. Accessed 3/17/24

[29] California Consumer Privacy Act. Cal. Civ. Code § 1798.100 (2018).

reminders, follow-up messages, and personalized health tips while safeguarding patient information. These CRMs connect marketing workflows with clinical systems and reduce the risk of unauthorized data sharing.

Despite these challenges, the digital landscape offers unprecedented opportunities for healthcare providers to personalize patient care and communication. Digital tools enable providers to tailor their interactions to individual patient needs and preferences, thereby enhancing the patient experience. This personalization can lead to improved patient outcomes and satisfaction, as well as increased efficiency in healthcare delivery.

However, achieving this balance requires a holistic approach that incorporates technology, policies, and a commitment to ethical data practices. By leveraging digital tools while ensuring the confidentiality, integrity, and availability of health information, healthcare providers can navigate the complexities of data privacy. This approach not only complies with regulatory requirements but also opens up new avenues for patient engagement and care personalization.[30] Conduct a privacy check of your website. Identify any third-party trackers and review whether your CRM and marketing tools meet HIPAA and CCPA standards.

With compliance in place, practices should define clear goals to guide their digital marketing efforts and measure performance.

Setting SMART Goals for Digital Success

Setting clear objectives is a foundational step in crafting a successful digital marketing strategy for healthcare providers. Identifying specific goals, such as increasing patient appointments or raising awareness of specialized services, provides a clear direction for your digital marketing efforts. These objectives should align with the broader business goals of your healthcare practice and be measurable, achievable, relevant, and time-bound (SMART).[31]

To measure the success of your digital marketing efforts, it's crucial first to identify the key performance indicators (KPIs) that align with your objectives. For example, suppose your goal is to increase patient appointments. In that case, your KPIs might include the number of online appointment bookings, website traffic to the appointment page, or calls to your practice originating from digital ads. Similarly, if raising awareness of specialized services is your goal, you might measure website page views for those services, engagement rates on related social media posts, or the number of new patients inquiring about those services.[32]

[30] https://thesiliconreview.com/2023/08/navigating-challanges-data-privacy-in-digital-age. Accessed 3/17/24.

[31] https://digitalmarketinginstitute.com/resources/lessons/digital-marketing-strategy_setting-objectives_2qgm. Accessed 3/17/24.

[32] https://www.proceedinnovative.com/blog/measure-digital-marketing-campaign/. Accessed 3/17/24.

By setting specific objectives and understanding how to measure their success, healthcare providers can more effectively navigate the digital landscape, optimize their marketing strategies, and achieve their desired outcomes. For example, build a 90-day plan with clear milestones. Assign tasks, monitor performance weekly, and adjust based on results.

Planning and Executing Your Digital Strategy

Creating a comprehensive digital marketing plan tailored to healthcare involves several critical steps. Initially, it's essential to understand the unique needs of healthcare providers and their clients. This understanding forms the basis for developing a strategy that addresses these specific requirements, ensuring that marketing efforts are both relevant and practical.[33]

1. **Identify your goals**: Clearly define what you aim to achieve with your digital marketing efforts. This could range from increasing patient appointments to enhancing the visibility of specialized services.[34]
2. **Understand your audience**: Knowing your target audience is crucial. Then you can tailor your strategy to patient preferences.[2]
3. **Select the right channels**: Choose digital channels that will most effectively reach your target audience. This could include search engine marketing (SEM), social media, email marketing, and other digital marketing channels.[35]
4. **Budget wisely:** Set a clear budget early. Most healthcare practices allocate between 2% and 10% of annual revenue to marketing, depending on growth goals and local competition [14]. New practices often spend more to build awareness, while established ones may shift focus to retention and patient engagement. Budget categories should include digital ad spend, content development, CRM tools, and analytics platforms [15].
5. **Develop content**: Create engaging and informative content that addresses the needs and concerns of your audience. Use this content to establish your expertise and build trust with potential patients.[36] Many healthcare practices now use generative AI tools, such as ChatGPT, Jasper, or Claude, to help draft blog posts, FAQs, emails, and video scripts. These tools can save time and support consistent content creation. However, they must be used carefully. They should never be used to process or store protected health information unless the tool is explicitly HIPAA-compliant. A qualified professional should always review

[33] https://comradeweb.com/blog/healthcare-digital-marketing/. Accessed 3/17/24.

[34] https://www.linkedin.com/pulse/6-healthcare-digital-marketing-strategies/?trk=public_post_main-feed-card_feed-article-content. Accessed 3/17/24.

[35] https://nogood.io/2023/10/18/healthcare-search-engine-marketing/. Accessed 3/17/24.

[36] https://www.kiwihealth.com/blog/digital-strategies-for-effective-medical-practice-marketing. Accessed 3/17/24.

outputs to ensure accuracy, tone, and alignment with practice guidelines. Generative AI is a support tool, not a replacement for clinical oversight or editorial judgement.

6. **Choose the right support:** Decide who will manage your digital marketing. A solo physician may benefit from hiring a freelance specialist for SEO or paid advertising. Multi-physician practices often turn to healthcare-specific marketing agencies or train in-house staff. Your decision should reflect your marketing complexity, oversight capacity, and available budget [16].
7. **Implement and monitor**: Launch your digital marketing campaign and closely monitor its performance. Utilize analytics tools to monitor your progress toward achieving your goals. Digital marketing requires ongoing refinement. A/B testing, that is, comparing two versions of an ad, email, or landing page, can reveal what resonates best with your target audience [17]. Even small changes in images, call-to-action language, or email subject lines may significantly increase conversions. Review performance metrics regularly and adjust accordingly [18].

The importance of continuously evaluating and adapting your strategies cannot be overstated. The digital landscape is dynamic, with patient behaviors and preferences evolving. Regularly reviewing performance data allows you to understand what's working and what isn't. This insight enables you to make informed decisions about adjusting your strategies to better meet the needs of your audience and achieve your marketing objectives more effectively.[37]

By following these steps and emphasizing the importance of continuous evaluation, healthcare providers can develop a digital marketing plan that not only reaches their target audience but also adapts to the changing digital environment, ensuring long-term success.

As digital marketing evolves, practices that monitor performance and refine their strategy will be best positioned to thrive in a competitive environment.

Conclusion: Adapting and Growing in the Digital Landscape

Digital marketing is now a clinical and business imperative. In today's environment, healthcare providers must do more than maintain a website or run occasional ads. Digital outreach supports patient education, promotes trust, and helps practices stay visible and responsive in a crowded marketplace.

As we look toward 2026, the most effective strategies rest on three pillars: automation, personalization, and trust. Automation lightens the administrative burden, ensures consistent messaging, and enables small practices to compete on a larger scale. Personalization tailors content and communication to each patient's

[37] https://growthnatives.com/blogs/digital-marketing/a-comprehensive-guide-to-healthcare-digital-marketing-agency/. Accessed 3/17/24.

needs, behaviors, or health status. Trust anchors every interaction. Patients want secure systems, accurate content, and transparency about how their data is handled.

Tools like behavior-based email campaigns, automated follow-up, and content libraries tied to clinical milestones let physicians engage patients outside the exam room. These efforts reinforce your role as a reliable guide, not just a provider. Done well, digital marketing supports your clinical mission and drives sustainable growth.

Digital outreach must match how patients live and communicate. In 2026 and beyond, your digital presence will shape first impressions, foster loyalty, and influence patient decisions. Now is the time to act with intention and lead.

References

1. Centers for Disease Control and Prevention. National health interview survey: adults who looked up health information online. 2023. https://www.cdc.gov/nchs/products/databriefs/db482.htm.
2. Healthgrades. 2023 Patient access journey report. 2023. Retrieved from https://practices.healthgrades.com/resources/patient-access-journey-report/.
3. Pew Research Center. Digital health behavior and information-seeking in the U.S. 2024. https://www.pewresearch.org.
4. Brown, S. Reputation management in healthcare: why it matters. 2023. https://www.growth.com/blog/232443/why-healthcare-reputation-management.
5. Accenture. Digital front door: accelerating the future of healthcare. 2021. https://www.accenture.com/us-en/insights/health/digital-front-door.
6. NexHealth. Patient experience platform for healthcare. 2023. https://www.nexhealth.com/product.
7. Doctible. Automated patient communication for modern practices. .2024. https://www.doctible.com/features.
8. Salesforce Health Cloud. Personalized patient engagement. 2023. https://www.salesforce.com/products/health-cloud/overview/.
9. Purcarea VL. The impact of marketing strategies in healthcare systems. J Med Life. 2019;12(2):93–6. https://doi.org/10.25122/jml-2019-1003. PMID: 31406509; PMCID: PMC6685306.
10. Moreira A, Alves C, Machado J, Santos MF. An overview of omnichannel interaction in health care services. Mayo Clin Proc Digit Health. 2023;1(2):77–93. https://doi.org/10.1016/j.mcpdig.2023.03.002. Epub 2023 Apr 3. PMID: 38013946; PMCID: PMC10069288.
11. U.S. Department of Health and Human Services. Use of online tracking technologies by HIPAA covered entities and business associates. 2022. https://www.hhs.gov/hipaa/for-professionals/privacy/guidance/hipaa-online-tracking/index.html.
12. STAT News. Hospitals' use of website tracking tools raises HIPAA concerns. 2023. https://www.statnews.com/2023/03/08/hospitals-meta-google-patient-data-hipaa/.
13. HIPAA Journal. Meta pixel and google analytics on healthcare websites may violate HIPAA. 2023. https://www.hipaajournal.com/meta-pixel-google-analytics-healthcare-websites-hipaa/.
14. HealthLeaders. How much should healthcare organizations spend on marketing?. 2022. https://www.healthleadersmedia.com/marketing/how-much-should-healthcare-organizations-spend-marketing.

15. Medical Group Management Association (MGMA). MGMA data dive: marketing costs in physician practices. 2021. https://www.mgma.com/data.
16. Forbes Agency Council. Should you hire in-house or outsource your marketing? Forbes. 2021. https://www.forbes.com/sites/forbesagencycouncil/2021/07/22/should-you-hire-in-house-or-outsource-your-marketing.
17. Google. A/B testing: Run experiments to improve performance. 2023. https://support.google.com/google-ads/answer/6381016.
18. HubSpot. The ultimate guide to A/B testing. 2022. https://blog.hubspot.com/marketing/how-to-do-a-b-testing.

Chapter 2
Creating a Practice's Digital Presence

As we discussed in Chap. 1, the digital landscape in healthcare is rapidly evolving, reshaping the way medical practices interact with patients. Technological advancements and changing patient expectations drive this transformation. Today, a digital presence is not just an option for healthcare providers; it's a necessity. It serves as the first point of contact between patients and medical practices, offering a platform for information exchange, service promotion, and patient engagement. To understand why this matters so much, it helps to look at the specific benefits a digital presence offers your practice.

The benefits of establishing an online presence are substantial. A well-crafted website can significantly extend a practice's reach, making its services accessible to a broader audience, far beyond its physical location. It also provides a platform for showcasing expertise, sharing patient testimonials, and highlighting specialized services, thereby enhancing the practice's credibility and trustworthiness. Moreover, an online presence facilitates patient engagement through educational content, health tips, and online appointment scheduling, improving patient satisfaction and loyalty. But content alone isn't enough. The way information is presented plays a critical role in keeping users engaged.

To keep visitors interested, you also need to present your information in a way that is easy to absorb. For example, incorporating engaging visuals, such as images, infographics, and videos, can significantly enhance user engagement, making complex medical information more digestible and appealing. Additionally, understanding the target audience and tailoring the content accordingly can drive engagement and build trust.[1]

The digital age has ushered in a new era for healthcare, where a strong online presence is indispensable for attracting and engaging patients. By leveraging digital platforms, medical practices can enhance their visibility, credibility, and patient

[1] https://chudovo.com/designing-medical-websites-that-drive-engagement-and-trust-23-best-practices-and-tips/. Accessed 3/2/24.

C. Quinn, *The Digital Practice*, https://doi.org/10.1007/978-3-032-07745-5_2

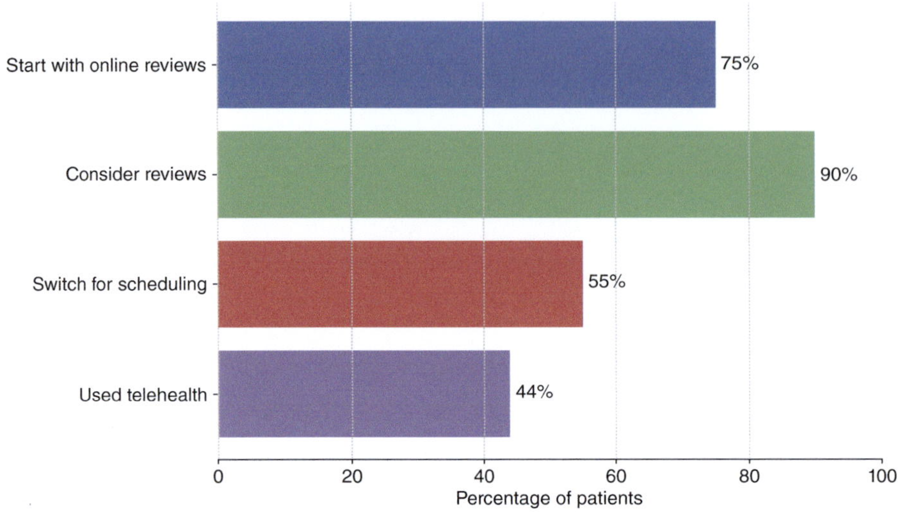

Fig. 2.1 Patient decision factors for selecting healthcare providers. (Quinn/ChatGPT)

engagement, ultimately leading to improved healthcare outcomes. The cornerstone of that digital strategy is a well-designed website, which serves as your digital front door.

Healthcare professionals need a website for several compelling reasons. First, a website enhances accessibility and patient communication. The reliance on traditional referral methods is declining. Patients are now forming their healthcare decisions well before visiting a doctor's office [1]. Recent data show that online reviews have become even more dominant. By 2024, 75% of patients will look at reviews as the first step in their search, and 90% of healthcare consumers say reviews influence their decisions.[2] These numbers reinforce why relying on outdated referral models is no longer sufficient. As shown in Fig. 2.1, online reviews and accurate information have become primary decision factors for patients when selecting a healthcare provider.

A practice can't depend on online directories provided by insurers to attract patients. Despite spending $2.76 billion annually to address the issue, Medicaid Services' online provider directories still suffer from inaccuracies. UnitedHealthcare's in-network directory is grossly inadequate,[3] and New York State's attorney general has stated that many insurers in her state publish inaccurate listings for providers or "ghost networks."[4] For healthcare practices aiming to enhance patient engagement and acquisition, prioritizing the improvement of their online presence is essential.

---

[2] https://b2b.healthgrades.com/insights/blog/online-reviews-impact-how-patients-select-hospitals-doctors/#:~:text=and%20are%20willing%20to%20do,for%20patient%20acquisition%20and%20outreach. Accessed 7/26/25.

[3] https://www.levernews.com/the-case-of-the-missing-health-care-providers/

[4] chrome-extension://efaidnbmnnnibpcajpcglclefindmkaj/https://ag.ny.gov/sites/default/files/reports/mental-health-report_0.pdf. Accessed 3/7/24.

Patients prioritize positive online reviews and relevant information when choosing a healthcare provider, showing a shift toward digital resources over traditional referrals. The availability of reviews and accurate information online, along with digital appointment scheduling, is a crucial factor influencing patients' decisions.

Despite the shift toward digital channels, consumers find the healthcare journey challenging due to a lack of sufficient online information about providers. Given this shift, practices need to think beyond waiting-room marketing and meet patients where they search, including online reviews, search engines, and social platforms. Because more patients start online, your website becomes your front door. That's why your website needs to be more than a digital brochure; it must offer both convenience and credibility.

Why a Practice Website?

A practice's website allows patients to access information about services, book appointments, and even consult with healthcare providers from home[5] (see Appendix 2.1 for a sample website structure). Convenience plays a major role in provider selection. According to a Healthgrades survey, 55% of patients said they would consider switching doctors to one who offers online scheduling.[4]

Your website shapes your reputation before you meet a single patient. From ease of booking to quality of content, every detail signals whether your practice values patient experience. But credibility isn't built on convenience alone.

A clear, well-maintained website also builds trust. Patients look for accurate information, transparency, and professionalism. When your site delivers on those expectations, it reassures visitors that your practice is competent, responsive, and trustworthy. This credibility helps attract new patients and retain current ones.[6]

A practice website offers an unparalleled opportunity to showcase your services and specialties to a wider audience. With the right design and content, your website can highlight the practice's unique offerings and expertise, making it stand out in a crowded healthcare market. This visibility is crucial for attracting patients seeking specific treatments or specialties.[7]

Figure 2.2 illustrates the core attributes of an effective healthcare website, highlighting the design and functional elements that build credibility and enhance user experience.

---

[5] https://townsquareinteractive.com/blog/medical-website-design-tips/. Accessed 3/2/24

[6] https://www.concretecms.com/about/blog/web-design/healthcare-website-design-best-practices-and-strategies. Accessed 3/2/24.

[7] https://myadvice.com/blog/website-design/5-ways-to-create-a-more-patient-friendly-website/. Accessed 3/2/24.

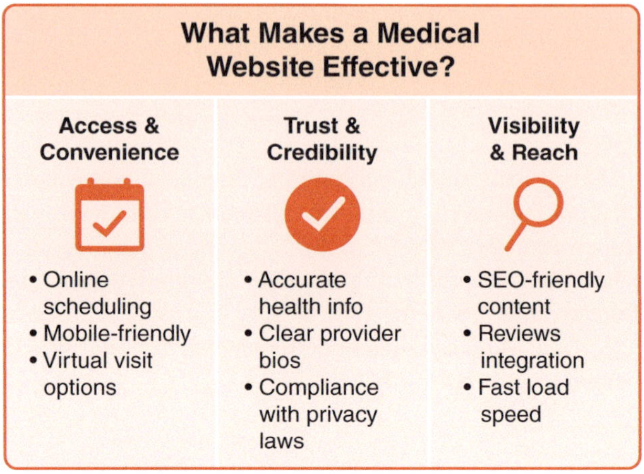

Fig. 2.2 Attributes of an effective website. (Image Quinn/ChatGPT)

Essential Components of a Healthcare Website

Creating a healthcare website that attracts and engages patients requires a strategic approach to its design and content. Here are the essential components:

Home Page

The home page serves as the gateway to your practice. It should introduce the practice and its mission, providing a clear and welcoming message to visitors. This page sets the tone for the user experience, highlighting the practice's values and commitment to patient care. A well-designed home page encourages users to explore the site further.[8]

Incorporating patient feedback during the development of digital healthcare tools is crucial for creating solutions that are both user-friendly and effective in meeting patients' needs. Engaging patients early in the design process helps identify and address their concerns, leading to tools that are more likely to be adopted and valued by the end users.

For example, research highlights the disconnect between the development of digital health tools and the patients they aim to serve. By involving patients from the outset, developers can bridge this gap, ensuring the tools are tailored to the actual needs of patients [2]. Another study emphasizes that patient-designed, patient-centered digital health tools are more likely to succeed. These tools bridge the gap between existing products and the populations they intend to serve, demonstrating the value of patient input in the design process [3].

---

[8] https://www.janbaskdigitaldesign.com/blogs/healthcare-website-design-trends/. Accessed 3/2/24.

Moreover, involving patients in the early stages of design is important for finding patient-focused solutions. This approach not only enhances the relevance of digital health interventions but also ensures their effectiveness in real-world settings.[9]

Clear Navigation

Patients visit your medical website with distinct objectives in mind, such as finding your location, learning about the procedures you provide, or seeking information on specific treatments.[10] To cater to these needs effectively, it's crucial to ensure that such information is readily available and easily accessible.

The first step toward achieving this goal involves designing a navigation menu that is both intuitive and straightforward. This menu should guide visitors effortlessly to the information they seek, thereby enhancing their overall experience on your site. Additionally, incorporating a search function can further streamline the process, enabling patients to quickly locate specific details without having to navigate through multiple pages.

Table 2.1 provides a quick reference guide to essential healthcare website components, their purposes, and practical tips for implementation.

Table 2.1 Quick reference: essential website components

Component	Purpose	Key tips
Home page	First impression; introduce mission	Clear welcome message; highlight values
Clear navigation	Guide visitors to content	Simple menu; include search bar
About us	Build trust and credibility	Share credentials; tell your story
Services offered	Explain treatments and specialties	Use patient-focused language; list options
Educational materials	Inform patients and peers	Provide articles, FAQs, health tips
Telehealth features	Enable virtual visits and scheduling	Promote convenience; easy booking links

[9] Gilbert R, Front. Digit. Health, 08 August 2022, Sec. Human Factors and Digital Health Volume 4 – 2022 | https://doi.org/10.3389/fdgth.2022.899976

[10] https://patientxagency.com/what-your-patients-want-to-see-on-your-medical-website/. Accessed 3/2/24.

Accessibility and Inclusivity

An effective healthcare website serves all patients, including those with visual, auditory, or motor impairments. Follow the Web Content Accessibility Guidelines (WCAG 2.1) by providing text alternatives for images, using high-contrast color schemes, and ensuring that every function is accessible through a keyboard.[11] (See Appendix 2.2 for a checklist of accessibility and compliance standards.) Captions on videos, descriptive alt text on graphics, and clear labels on form fields help users who rely on screen readers. A responsive layout that works with assistive technologies not only meets legal requirements but also signals respect for every visitor. If your site is accessible, you reach everyone, not just those with perfect eyesight and hearing.

See Appendix 2.1 at the end of this chapter for a Web Content Accessibility Guideline checklist.

About Us

The "About Us" page offers background information on healthcare professionals and staff, establishing trust and credibility. It's an opportunity to share qualifications, experience, and the personal stories that connect staff with patients. This page should articulate who you are, why you do the work that you do, and what sets your practice apart.[12] It should also humanize you. Share why you chose medicine or how your practice has evolved.

Services Offered

Detailed descriptions of services and treatments available are crucial. This section should be patient-centric, explaining how each service addresses patients' needs and improves their health outcomes. It is essential to avoid medical jargon, as clear and concise medical information helps patients understand what to expect and how to prepare for their visit.[13]

[11] https://www.wcag.com/. Accessed 7/27/25.

[12] https://healthcaresuccess.com/blog/healthcare-marketing/how-to-write-an-about-us-page-for-a-healthcare-business-with-examples.html. Accessed 3/2/24.

[13] https://www.practiceedge.com.au/16-essential-elements-for-a-successful-healthcare-website/. Accessed 3/2/24.

Educational Materials, FAQs, and Health Tips

Providing educational resources on a healthcare website is essential for informing other healthcare providers about your practice's specialties and interests, as well as empowering patients with the knowledge they need to manage their health effectively.

Information provided on your practice's websites can enhance patients' understanding, skills, and involvement in making health-related decisions [4]. This online resource complements the constrained time of face-to-face doctor-patient consultations, improving their effectiveness [5]. It allows patients to seek answers to questions they may have forgotten to ask during their visit or to inquire about topics they might find too sensitive to discuss face-to-face. This way, patients can obtain the information they need comfortably and privately from their own homes.

Creating a medical content strategy presents unique challenges due to the diverse needs and expectations of its audience. This complexity is rooted in the necessity to balance technical accuracy with accessibility, ensuring that content is both medically reliable and understandable to nonprofessionals.

The first challenge in content creation is understanding the distinct needs of doctors and patients who visit the website. Healthcare professionals seek detailed, research-based information that supports their clinical decisions and professional development. In contrast, patients seek information that is easy to understand and directly applicable to their specific health concerns.[14] Bridging this gap requires a nuanced approach to content creation, where information is presented in multiple layers or formats tailored to each audience segment.

A prime example of this approach in action is a healthcare provider who implemented a dynamic content strategy on their website. By analyzing patient data, including location, browsing history, and online behavior, they were able to adjust the content displayed on the website to offer personalized health information and interactive tools that resonated with patients at various stages of their journey.[15]

Medical knowledge evolves quickly, so you need a process for keeping your site up to date and trustworthy. Every educational piece should be dated, assigned to a medical reviewer in the practice, and checked at least twice a year. That reviewer verifies each article, FAQ, and health tip for medical accuracy, updates any sections to reflect current research, and replaces outdated information. By clearly showing when content was last reviewed and who reviewed it, you signal to readers and search engines alike that your information is credible and fresh. This rigor protects patients from outdated guidance and boosts your reputation as a reliable source.

The content on your website should try to answer common patient questions. Incorporating a Frequently Asked Questions (FAQ) section can address these common concerns and streamline patient communication. Health tips, grounded in

---

[14] https://wellbe.me/how-to-enhance-the-patient-experience-at-every-touchpoint/

[15] https://silverlinecrm.com/blog/healthcare/personalization-in-healthcare-patient-experience-use-cases/

evidence-based practices, can encourage patients to adopt healthier lifestyles. The presentation of the information on your website is also important. The average reader should be able to understand all the content on your website. You should avoid medical jargon and use words and terms the patient will understand (e.g., high blood pressure rather than hypertension).

Educational materials should cover a wide range of topics relevant to the practice's specialties (e.g., diseases treated, procedures performed) and should be accessible in various formats (e.g., articles, videos, infographics) to accommodate different learning preferences. The CDC offers a wealth of materials that can support your patient education efforts, including reports, guidelines, and recommendations.[16]

Contact Information: Easy-to-Find Contact Details and Online Appointment Booking

Ensuring that contact information is easily accessible and that patients can book appointments online is a critical component of a healthcare website. This convenience enhances the patient experience and improves accessibility to care. Tools like Calendly[17] and Acuity Scheduling[18] offer seamless integration for online appointment booking, enabling patients to schedule visits without needing to call the office. This not only saves time for both patients and staff but also reduces the barriers to securing appointments. Including clear, visible contact details—such as phone numbers, email addresses, and physical location maps—ensures that patients can easily reach out with questions or for emergency purposes.

A healthcare website must be more than just a digital brochure. It should be a resource that educates, engages, and reassures current and prospective patients. By focusing on these essential components, you create a platform that effectively communicates your practice's mission, values, and services.

Features to Enhance User Experience

To attract and engage patients effectively, certain website features stand out for enhancing user experience.

First, a responsive, flexible design is nonnegotiable. Flexible design in a website refers to the approach of creating web pages that can adjust and adapt to various screen sizes, platforms, and orientations. This design philosophy prioritizes the

---

[16] https://www.cdc.gov/dhdsp/materials_for_professionals.htm. Accessed 3/2/24.

[17] https://calendly.com/. Accessed 3/2/24.

[18] https://acuityscheduling.com/. Accessed 3/2/24.

user's environment, ensuring that the website provides an optimal viewing and interaction experience across all devices, from desktop computers to mobile phones and tablets.

By employing responsive web design techniques, designers and developers can make web pages that dynamically change layout, size, and content arrangement based on the device's screen dimensions and capabilities. This adaptability not only enhances user experience but also supports better accessibility and usability of the website across different devices.[19]

With most of the Internet traffic coming from mobile and tablet devices, a website that adapts seamlessly to any screen size ensures that patients can access information, book appointments, and navigate services effortlessly, regardless of the device they use.[20]

Second, secure patient portals are a cornerstone of modern medical websites. These portals offer patients a private gateway to their personal health information, allowing them to view test results, schedule appointments, and communicate with their healthcare providers securely. This feature not only streamlines administrative processes but also empowers patients in their healthcare journey, making it a critical component of patient-centered care.[21]

Lastly, a blog section dedicated to the latest health news, research, and practice updates serves multiple purposes. It educates patients, keeping them informed about the latest in health and wellness. It also positions the practice as a thought leader in the medical field, building trust and credibility. Regularly updated content boosts SEO, increasing the website's visibility and attracting more visitors.[22]

Regularly Test and Update Your Website

Regular testing and updating of a medical website are important for several reasons. Firstly, it ensures the accuracy and relevance of the content, which is vital in the healthcare domain where information can directly impact patient care and decision-making. Regular updates reflect the latest health guidelines, research findings, and treatments, thereby maintaining the website's credibility and trustworthiness.[23]

---

[19] https://pixetic.com/blog/responsive-web-design-how-to-make-your-website-flexible/. Accessed 3/3/24.

[20] https://www.linkedin.com/pulse/13-must-have-features-medical-practice-website-netwalker-sng/. Accessed 3/3/24.

[21] https://www.growthplug.com/blog/6-must-have-features-in-your-healthcare-practice-website. Accessed 3/3/24.

[22] https://healthcaresuccess.com/blog/healthcare-advertising-agency/10-ui-ux-principles-to-enhance-your-healthcare-website-experience.html. Accessed 3/3/24.

[23] https://ahamediagroup.com/blog/how-often-should-you-refresh-your-healthcare-content/. Accessed 3/3/24.

You need more than a glossy design; you need a site that works reliably and keeps your patients' data safe. Site testing encompasses everything from the ease of booking an appointment for patients to the effectiveness of your security measures in protecting health information. In a small practice, your internal IT team might handle usability testing by watching staff or a few patients navigate the site and noting where they stumble. Larger practices often hire outside security consultants to perform vulnerability scans and penetration tests that look for flaws in log-in systems, encryption, and data storage.

The testing should include all interactive elements: online forms, appointment schedulers, chat tools, and anything that collects patient information.

Security tests check for common web attacks. Instead of diving into technical jargon, such as penetration testing, SQL injection, and cross-site scripting, think of website security testing as hiring someone to shake the doors and windows of your digital office to make sure they're locked. The testers look for the online equivalent of an open window that could allow a stranger to view patient data.

Usability testing is like watching a new patient try to find your waiting room. If they get lost because the signs aren't clear or the mobile site hides the directions, you'll know it's time to fix those issues.

Neglecting these tests can have serious consequences. Bugs in an appointment form can cause lost bookings or duplicate appointments. Unpatched security vulnerabilities can expose patient data, potentially leading to identity theft or triggering fines and lawsuits under HIPAA. Even seemingly minor issues, like a broken link to privacy policies, can erode trust. Regular testing and prompt repairs keep your site running smoothly, protect patient privacy, and demonstrate your commitment to quality care.

Regular testing like this helps maintain high security and compliance with health information privacy standards, such as HIPAA in the United States.[24]

Moreover, testing for functionality and user experience ensures that the website remains user-friendly, accessible, and efficient in helping users find the information they need or complete actions such as booking appointments. This enhances patient satisfaction and engagement, which are key to a healthcare provider's success online.

Encourage Patient Reviews

In the digital age, where many patients first encounter healthcare providers online, reviews can significantly influence decision-making. Here's a deeper dive into managing and leveraging patient reviews:

Encouraging patients to leave reviews is the first step in harnessing the power of patient feedback. You can do this by sending follow-up emails after appointments, displaying signs in the office, or verbally asking satisfied patients to share

[24] https://testsigma.com/blog/healthcare-domain-testing/. Accessed 3/3/24.

their experiences online. It's important to make the process as easy as possible, providing direct links to review platforms.

You should regularly monitor reviews across platforms like Google, Healthgrades, and Yelp. This allows you to stay informed about patient sentiment. Responding to reviews, both positive and negative, demonstrates that the practice values feedback and is committed to patient satisfaction. Responses should be professional, HIPAA-compliant, and personalized when possible, addressing any specific concerns raised.

Positive reviews can be powerful marketing tools. Highlighting patient testimonials on your website, in social media posts, or in marketing materials can build trust with prospective patients. It showcases real-life success stories and the positive impact of your practice on patients' lives.

Negative reviews offer an opportunity for improvement.

By addressing concerns raised in such reviews, practices can demonstrate their commitment to quality care and patient satisfaction. It's also an opportunity to rectify any misunderstandings and to invite the reviewer to discuss their experience further offline, which can sometimes lead to changing a negative review into a positive one.

Search engines value fresh, original content, and patient reviews are a form of content that can help improve a practice's search engine rankings. Additionally, reviews often contain keywords relevant to the practice's services, which can further enhance SEO efforts.

To efficiently manage reviews, consider using a review management system or software, such as Birdeye, Podium, or Reputation.com. These tools can help aggregate reviews from multiple platforms, alert you to new reviews, and streamline the response process. This ensures that no review goes unnoticed and that the practice maintains an active, engaging online presence.

What to Exclude from Your Website

When managing a medical website, it's crucial to maintain professionalism and comply with legal and ethical standards. Here are key elements to exclude:

- Personal health information without consent: Never display personal health information (PHI) without explicit consent from the patient. This is a fundamental aspect of HIPAA compliance, protecting patient privacy and confidentiality.[25]
- Unverified medical advice or claims: Avoid publishing medical advice or health claims that are not backed by scientific research or approved by relevant health

[25] https://www.practicebuilders.com/blog/is-your-medical-practice-website-hipaa-compliant/. Accessed 3/3/24.

authorities. This ensures the information provided is reliable and trustworthy, safeguarding your practice against misinformation and potential legal issues.[26]

- Third-party advertisements that may compromise professionalism: Refrain from including third-party advertisements, especially those unrelated to medical services, as they can detract from the professionalism of your site and potentially mislead patients. Ensuring the content aligns with the medical practice's values and mission is essential for maintaining trust and credibility.[27]

Maintaining Compliance and Patient Privacy

Healthcare providers build patient trust by demonstrating that their data is secure and that breaches are effectively managed to prevent sensitive information from being compromised.[28]

One way of doing this is by maintaining compliance with the Health Insurance Portability and Accountability Act (HIPAA). HIPAA sets the standard for protecting sensitive patient data such as secure transmission and storage of patient data and implementing safeguards against data tampering.[29,30] See Appendix 2.2 for a HIPAA compliance checklist for your website.

By implementing multilayered security measures, including encryption, secure patient portals, and regular risk assessments and audits,[31] a practice can protect its patient information from unauthorized access.[32] These measures ensure that patient data is not only secure but also accessible only to authorized personnel.[33]

HTTPS (HyperText Transfer Protocol Secure) is the secure version of HTTP, the protocol that governs how data is transmitted between a website and a user's browser. For medical practices, using HTTPS is essential for protecting sensitive information, such as patient intake forms, appointment requests, or log-in credentials. By encrypting data in transit, HTTPS helps prevent unauthorized access and supports HIPAA compliance. In addition to improving security, it also enhances

[26] https://www.seomedical.com/blog/medical-web-copywriting/medical-practice-content-update.htm. Accessed 3/3/24.

[27] https://megadigital.ai/en/blog/restricted-medical-content-google-ads/. Accessed 3/3/24.

[28] https://www.linkedin.com/pulse/data-security-healthcare-ensuring-patient-privacy-digital-shariful/

[29] https://redblink.com/hipaa-compliant-healthcare-website-design-best-practices/. Accessed 3/3/24.

[30] https://www.clarity-ventures.com/services/hipaa-compliant-websites. Accessed 3/3/24.

[31] https://www.itsasap.com/blog/hipaa-security-best-practices. Accessed 3/3/24.

[32] https://www.clarity-ventures.com/hipaa-guidelines/hipaa-compliance-requirements-and-best-practices. Accessed 3/3/24.

[33] https://fastercapital.com/content/Redacted-Medical-Records%2D%2DProtecting-Patient-Privacy-in-the-Digital-Age.html#:~:text=Ensuring%20patient%20privacy%20in%20the%20digital%20era%20requires%20a%20multi,regular%20risk%20assessments%20and%20audits

trust, as most modern browsers display a padlock icon for HTTPS-enabled sites, signaling safety to visitors. Search engines like Google also favor HTTPS sites in their rankings, making it a simple but important step in improving visibility and credibility online. If your website address begins with "https://," it is secure; if not, ask your web developer or hosting provider to install an SSL certificate.

Securing the website with an SSL certificate is another step toward creating a HIPAA-compliant website.[34] An SSL certificate is a digital certificate that authenticates the identity of a website and enables an encrypted connection.[35] SSL stands for Secure Sockets Layer, a security protocol that creates a secure link between a web server and a browser. This encrypted link ensures that all data passed between the web server and browsers remain private and integral, protecting against eavesdropping, tampering, and message forgery.

SSL certificates are essential for websites to use HTTPS. They are issued by Certificate Authorities (CAs) to organizations and their domains/websites after verifying the organization's identity, thereby increasing user confidence by showing that an industry-recognized authority has validated the website's identity.[36,37]

HIPAA rules apply to social media just as they do to websites. This includes any content from your practice that goes out on sites like TikTok, LinkedIn, or YouTube. The *HIPAA Journal* warns that no social media post should include protected health information (PHI).[38]

Another important way of protecting patient data is by training employees on HIPAA regulations, ensuring that all staff members understand their roles in maintaining patient privacy.[39]

Recent enforcement cases illustrate the risk: a dental practice (Elite Dental Associates in Dallas, Tx) was fined $10,000 in 2019 after staff impermissibly posted PHI on a review site, and a nursing assistant lost her job and served jail time in 2016 for sharing a patient video online.[40]

In another case that occurred in 2016, a nursing assistant at Parkside Manor in Kenosha, WI, lost her job and was sentenced to 30 days of jail time after taking and sharing a video of a mostly naked Alzheimer's patient on Snapchat without consent. The incident was reported, and she ultimately pled guilty to capturing an image of nudity without consent, a misdemeanor offense.

To protect your practice and your patients, develop a written social media policy, educate all employees on HIPAA requirements, and monitor your accounts for compliance.

---

[34] https://www.hipaaguide.net/hipaa-compliant-website/. Accessed 3/4/24.

[35] https://usa.kaspersky.com/resource-center/definitions/what-is-a-ssl-certificate. Accessed 3/4/24.

[36] https://www.cloudflare.com/learning/ssl/what-is-an-ssl-certificate/. Accessed 3/4/24.

[37] https://www.digicert.com/what-is-an-ssl-certificate. Accessed 3/4/24.

[38] https://www.hipaajournal.com/hipaa-social-media/#:~:text=HIPAA%20And%20Social%20Media%20Guidelines. 7/25/25.

[39] https://www.itsasap.com/blog/hipaa-security-best-practices.Accessed 3/3/24.

[40] https://www.hhs.gov/hipaa/for-professionals/compliance-enforcement/agreements/elite/index.html. Accessed 7/26/25.

Figure 2.3 summarizes HIPAA compliance dos and don'ts for healthcare websites, serving as a visual checklist for safeguarding patient information.

This section in HIPAA already underscores the importance of safeguarding protected health information. Still, recent guidance from the US Department of Health and Human Services adds a new layer of urgency.

In 2024, the HHS warned that cookies, pixels, and other tracking tools on healthcare websites can inadvertently capture PHI. Marketing or analytics tools, such as pixels, track visitor behavior on websites. When these tools record names, appointment details, or conditions searched, they collect protected health information.

Some tools send that information back to the vendor to build user profiles or target ads. For example, if a patient fills out a "request an appointment" form and a script linked to the form transmits the patient's name and reason for the visit to an analytics company, that company can match those details with other online activity. This practice discloses the patient's health information without consent. HIPAA prohibits this because it poses a risk of identity theft and discrimination. An insurance company or employer could gain access to sensitive health data through these profiles. To avoid this, configure your analytics tools to collect only anonymous, aggregated data, and never capture specific patient details.

The HHS bulletin reminds providers that sharing PHI with tracking vendors for marketing purposes without a patient's authorization is impermissible. Violations of this rule can expose individuals to identity theft, discrimination, and other harms.

To comply, configure analytics and tracking tools so they do not collect or transmit any data that could identify an individual. This means stripping out form fields, appointment details, or IP addresses before they reach third-party analytics. Work with vendors willing to sign Business Associate Agreements, and agree to the "minimum necessary" standard. Audit scripts regularly, disable remarketing on pages related to specific conditions or services, and avoid linking site activity to any unique identifiers. These steps enable you to measure site performance and enhance user experience without compromising the confidentiality, integrity, and security of PHI.

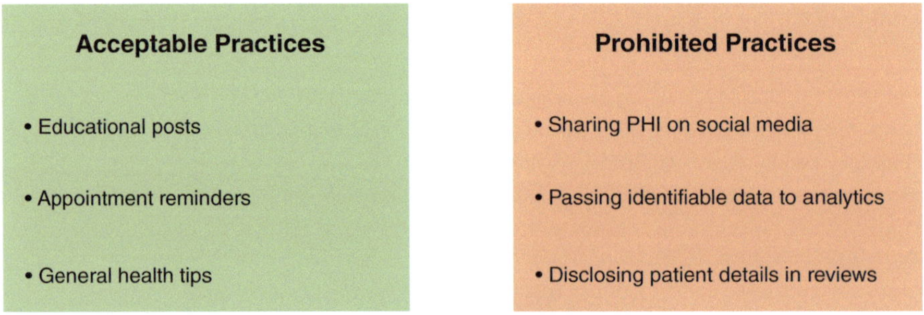

Fig. 2.3 HIPAA compliance dos and don'ts. (Image Quinn/ChatGPT)

Sidebar: Understanding Cookies, Pixels, and Patient Privacy

What Are Cookies?

Cookies are small text files stored on a user's device when they visit a website. They remember user preferences, such as log-in information or previously viewed pages. Some cookies also track behavior across websites for marketing purposes, like showing ads based on your browsing history.

What Are Tracking Pixels?

Tracking pixels are tiny, invisible images embedded in a website, or when the page loads or the email is opened, the pixel sends information (like what page was viewed and when) to third-party servers, such as Google or Meta (Facebook). This helps marketers analyze user behavior and measure ad performance.

Why Should Physicians Be Concerned?

In healthcare, some tracking tools may collect or transmit protected health information (PHI) without the user's consent, especially when placed on patient portals, appointment request pages, or symptom checkers. For example, if a patient searches your site for information on depression and books an appointment, tracking tools could unintentionally send this data to external marketing platforms. This could violate HIPAA.

The Risk

In December 2022, the US Department of Health and Human Services issued a bulletin clarifying that sharing PHI with third-party tracking vendors without explicit patient authorization is a HIPAA violation. Such unauthorized disclosures could expose patients to risks, such as identity theft or discrimination, and expose your practice to legal penalties.

What to Do

Avoid embedding tracking pixels or cookies on pages that collect or display sensitive health information.

Collaborate with web developers who are familiar with HIPAA-compliant marketing practices.

Use HIPAA-compliant analytics tools (e.g., from providers who sign a Business Associate Agreement).

Review all plug-ins and third-party scripts on your site on a regular basis.

Protecting patient data must be a top priority, even in your marketing strategy.

Lastly, the use of patient portals is recommended for secure communication between patients and healthcare providers. These portals are designed to safely send and receive messages, thereby enhancing the privacy and security of digital health interactions.[41]

By adhering to these guidelines and regularly updating their security measures, healthcare providers can ensure they maintain compliance and effectively protect patient privacy.

After you build a compliant, informative site, you need a plan to drive visitors to it. We'll see how to do that in the next section.

Marketing Your Practice Through Your Website

Marketing your medical practice through your website requires a strategic approach. One essential element is search engine optimization (SEO). SEO is the process of improving your website so it appears higher in Google search results when patients look for medical services. For example, if your practice specializes in dermatology, strong SEO helps your website show up when someone searches for "skin doctor near me." Effective medical SEO involves using relevant keywords, ensuring fast page load times, and making your site easy to view on mobile devices. These optimizations increase your visibility online and help attract more patients who are actively looking for care.[42, 43]

Next, leveraging social media is an effective way to drive traffic to your website. Platforms like Facebook, X (formerly Twitter), and Instagram enable you to share educational materials, patient testimonials, and health tips. This not only engages your current patient base but also attracts new patients by showcasing your expertise and care philosophy.[44, 45]

Finally, measuring the effectiveness of your online presence is essential. Tools like Google Analytics can track website traffic, user engagement, and conversion rates. Monitoring these metrics helps you understand what's working and where improvements are needed, ensuring your digital marketing efforts yield the best results for your practice.[46]

By focusing on these key areas, you can create a robust online marketing strategy that enhances your practice's visibility, engages patients, and, ultimately, drives more appointments.

---

41 https://telehealth.hhs.gov/patients/additional-resources/data-privacy

42 https://www.semrush.com/blog/healthcare-seo/. Accessed 3/4/24.

43 https://comradeweb.com/blog/medical-seo-tips-to-improve-seo-for-medical-websites/. Accessed 3/4/24.

44 https://blog.hootsuite.com/social-media-health-care/. Accessed 3/4/24.

45 https://www.webfx.com/industries/health/hospitals-healthcare/social-media/. Accessed 3/4/24.

46 https://www.getdeardoc.com/post/healthcare-marketing-metrics. Accessed 3/4/24.

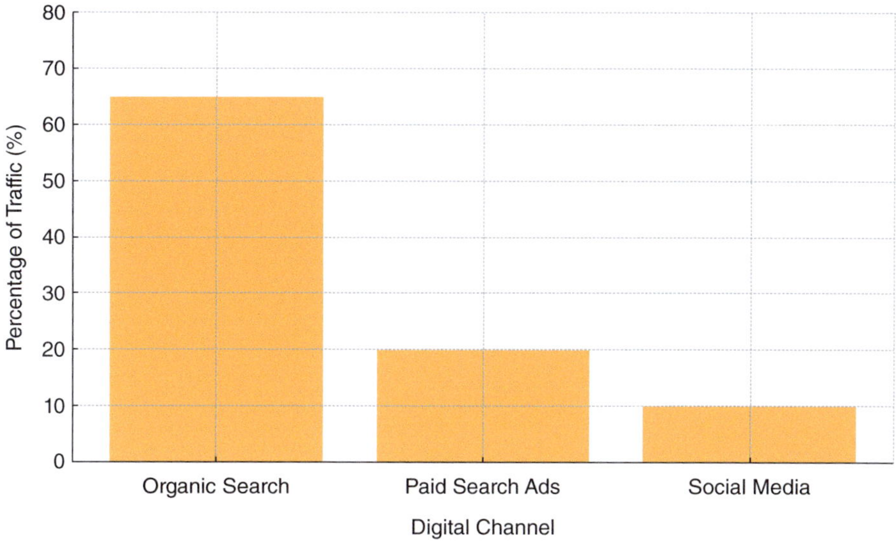

Fig. 2.4 Patient website traffic sources. (Image Quinn/ChatGPT)

Understanding Traffic Sources: What the Numbers Say

Data from the 2024 Healthcare Internet Conference reveals how patients find healthcare websites.[47] The majority, approximately 65%, arrive through organic search, indicating the critical role of search engine optimization (SEO) in medical marketing. Paid search ads account for about 20% of traffic, offering targeted visibility for specific services. Social media contributes around 10%, suggesting it serves better as a supporting rather than primary channel. This distribution highlights the importance of prioritizing high-quality, search-optimized content while utilizing ads and social platforms to supplement outreach efforts.

Figure 2.4 shows the primary sources of traffic to healthcare websites, emphasizing the importance of organic search in attracting patients.

Telehealth and Digital Advertising

Patients no longer see telehealth as an experiment. A 2024 survey found that 44% of respondents had a virtual visit in the previous year, and 94% of those patients were willing to do it again because it saved time and reduced waiting.[48] Virtual visits now

[47] https://www.invoca.com/blog/healthcare-marketing-statistics. Accessed 7/27/25.

[48] https://www.deloitte.com/us/en/insights/industry/health-care/virtual-health-consumer-demand-and-availability.html. Accessed 7/26/25.

account for nearly 46% of all medical appointments, up from 1% before 2020.[49] This surge in adoption means that telehealth capabilities can no longer be treated as an optional service.

Integrating telehealth into your digital advertising strategy is not about hopping on a trend; it reflects the way patients already engage with healthcare. By addressing real-world needs, speed, convenience, and trust, you can position your practice as a go-to option for modern patients and stand out in a crowded virtual marketplace.

Sidebar: What Patients Value in Virtual Visits
When it comes to telehealth, patients prioritize convenience and efficiency. Key expectations include the following:

- **Short wait times**: Delays lead to dissatisfaction, even in virtual settings. Patients expect to be seen within minutes of their scheduled time.
- **Easy booking**: A streamlined, intuitive appointment process, especially with mobile access, is essential.
- **Clear instructions**: Patients want to know exactly how to log in, what to expect, and how to follow up.
- **Privacy**: Assurance that their information is secure builds trust.
- **Follow-up support**: Whether it's a summary of the visit or a prompt response to post-visit questions, continuity matters.

Meeting these expectations can improve patient satisfaction, retention, and reviews.

Without measurement, all your marketing is guesswork. In the next section, we'll examine the role of analytics in marketing your practice.

Analytics

Digital analytics plays a crucial role in understanding the effectiveness of online content, including marketing campaigns, social media, and websites. By analyzing data on how users interact with your digital platforms, you can pinpoint areas where your strategies may not be reaching their intended audience or achieving desired outcomes. This process of identifying gaps is essential for making informed decisions on where to allocate resources more effectively to enhance performance.

---

[49] https://direction.com/telehealth-digital-marketing-strategies/#:~:text=The%20telehealth%20boom%20created%20massive,for%20the%20same%20patients%20online. Accessed 7/26/25.

For instance, if digital analytics reveal that your webpage has a high bounce rate, meaning that visitors leave the page quickly after arriving, this indicates a gap in engaging the audience.

Similarly, if an analysis of social media engagement shows low interaction rates with certain types of posts, this highlights an area for improvement.

Once these gaps are identified, you can develop targeted interventions to address these issues.

This might involve redesigning webpages to make them more engaging, tailoring content to better meet the interests of the audience or optimizing marketing messages to increase their appeal. The goal of these interventions is to improve user engagement, enhance satisfaction, and ultimately drive better business outcomes.

An example of customizing a website to enhance patient engagement could be that new visitors to the site were greeted with general health information and an interactive symptom checker tool while returning visitors saw content tailored to their previous searches or conditions of interest.

This strategy led to a marked increase in patient engagement, as measured by time spent on the site and conversion rates for appointment bookings.

Challenges

Digital healthcare, while transformative, faces significant challenges that can hinder its effectiveness and accessibility.

Firstly, the digital divide remains a pressing issue. This gap between those who have access to digital technologies and those who do not can exacerbate health inequalities. For example, rural and underserved communities often lack the necessary infrastructure for reliable Internet access, limiting their ability to benefit from digital health services [6].

Technology malfunctions also pose a challenge. Software glitches, system downtime, and hardware failures can disrupt healthcare delivery, potentially leading to delays in treatment or errors in patient care. Such technical issues underscore the need for robust, reliable digital health platforms [7].

Moreover, the impersonal nature of some digital interactions can affect the patient-provider relationship. While digital tools offer convenience and efficiency, they can sometimes lack the personal touch that face-to-face interactions provide, potentially impacting patient satisfaction and trust.[50]

Addressing these challenges requires concerted efforts from policymakers, healthcare providers, and technology developers. By acknowledging and tackling these issues, the healthcare sector can ensure that digital health tools are more inclusive, reliable, and patient-centered.

---

[50] Cummins N, Schuller B, Front. Digit. Health, 07 December 2020, Sec. Connected Health, Volume 2 – 2020 | https://doi.org/10.3389/fdgth.2020.536203

If This All Seems a Little Too Much …

If managing a website seems overwhelming for the average health practitioner, there are several viable alternatives to consider. These alternatives can help maintain an online presence without the complexities of direct website management:

1. **Medical website builders**: Utilizing medical website builders specifically designed for the healthcare sector can simplify the process. These platforms often come with templates and tools tailored to healthcare needs, making it easier to create and manage a professional website without needing advanced technical skills.[51,52] These are ideal for solo providers or small group practices that want a fast, affordable, and HIPAA-compliant website without needing to hire a developer. Examples include platforms like PatientPop or Wix for Healthcare. They provide healthcare-specific templates, SEO support, and patient engagement tools.
2. **Practice management software**: For those looking for a more comprehensive solution, practice management software offers a range of features beyond just website management. These platforms can cater to different healthcare fields, offering patient scheduling, billing, and even telehealth services, thereby reducing the need for a separate website.[53,54] It is best for midsize practices that need an all-in-one platform for scheduling, billing, and communications. Platforms like Kareo, SimplePractice, or Athenahealth often include built-in web portals, reducing the need for a separate website and centralizing operations.
3. **Clinical management system alternatives**: Exploring alternatives to traditional clinical management systems can also provide integrated solutions that include patient engagement tools, online booking, and more, potentially reducing the reliance on a standalone website for patient interaction.[55] This option is suitable for practices looking to modernize legacy systems or add features like telehealth, online booking, or mobile patient communication. These tools (e.g., DrChrono, Elation Health) often integrate website-like functions with clinical workflows, making them a strong choice for tech-forward clinics or those growing rapidly.

These alternatives provide streamlined, user-friendly options for health practitioners to maintain an online presence and manage patient interactions without the need for in-depth website management. You should choose an option that fits your goals.

---

[51] https://healthcarebusinessclub.com/articles/business-supplier/business-management/9-best-medical-website-builders-for-the-healthcare-sector/#google_vignette. Accessed 3/4/24.

[52] https://morweb.org/post/medical-website-builders. Accessed 3/4/24.

[53] https://practice.do/blog/12-practice-better-alternatives. Accessed 3/4/24.

[54] https://www.medesk.net/en/alternative-health-practice-management-software/. Accessed 3/4/24.

[55] https://www.g2.com/products/clinical-management-system/competitors/alternatives. Accessed 3/4/24.

Conclusion

Establishing a strong online presence is no longer optional for medical practices; it's a necessity. The digital age has transformed how patients seek and engage with healthcare providers. A robust online presence, encompassing a user-friendly website, active social media engagement, and effective digital marketing strategies, is vital for attracting new patients, establishing credibility, and enhancing patient-provider relationships. As we've seen, digital and social media are essential tools for modern physicians to relate to their patients and expand their reach. By embracing these digital avenues, healthcare professionals can not only meet but exceed patient expectations, ensuring their practice remains competitive and relevant in today's healthcare landscape. By investing in your online presence, you not only meet patients where they are, but you also stand out in a crowded market.

Appendixes

Appendix 2.1: Web Accessibility Compliance Checklist

This checklist is based on the Web Content Accessibility Guidelines (WCAG) 2.1 Level AA. Use it to evaluate and improve the accessibility of your medical practice's website.

☐ Use alt text for every meaningful image. Screen readers rely on these descriptions to convey image content to visually impaired users.
☐ Choose high-contrast color combinations. Ensure at least a 4.5:1 contrast ratio for text to enhance readability for users with low vision.
☐ Enable keyboard-only navigation. All interactive elements (menus, buttons, forms) must be operable using the Tab key and Enter/Space.
☐ Provide captions and transcripts for all video and audio content. This ensures access for deaf and hard-of-hearing users.
☐ Label all form fields clearly with visible text and provide plain-language error messages for invalid entries.
☐ Allow text resizing up to 200% without breaking the layout or hiding content.
☐ Write descriptive link text (e.g., "Schedule an appointment" instead of "Click here") to support screen reader navigation.
☐ Use accessibility tools to test your site. Tools like WAVE, Axe, and screen readers help evaluate compliance and identify issues.
☐ Set the language for each webpage using appropriate language tags (e.g., <html lang="en">).
☐ Ensure mobile responsiveness so that content reflows without horizontal scrolling.
☐ Provide controls to pause or stop any auto-advancing content (e.g., slideshows or alerts).

Checking off these items will make your site more accessible for everyone and help you stay compliant with accessibility regulations.

Appendix 2.2: HIPAA Compliance Checklist (Physician Website Version)

Security and Encryption

[ ] The website uses HTTPS with a valid SSL certificate.
[ ] All online forms are encrypted in transit using secure protocols.
[ ] The hosting provider offers HIPAA-compliant server infrastructure.
[ ] Uploaded files such as medical records are encrypted at rest.

Data Collection and Communication

[ ] Online forms request only necessary health information.
[ ] Each form includes a HIPAA disclaimer or patient consent checkbox.
[ ] Emails generated by the website avoid including protected health information unless they are encrypted.
[ ] Any services used to build or manage forms have a signed Business Associate Agreement (BAA).

Third-Party Tools and Tracking

[ ] No PHI is shared with third-party marketing tools or trackers such as Google Analytics or Meta Pixel.
[ ] All third-party vendors who may access PHI have signed BAAs.
[ ] Live chat or chatbot tools are either HIPAA-compliant or not used.
[ ] A cookie consent notice is included if the website uses tracking technologies.

Data Access and Storage

[ ] Access to PHI collected through the website is limited to authorized staff.
[ ] Access is logged, and audit records are maintained.
[ ] Secure, encrypted backups are performed regularly.
[ ] A data retention and deletion policy is documented and followed.

Policies, Notices, and Transparency

[ ] A privacy policy is published that explains how patient data is collected, stored, and used.
[ ] A Notice of Privacy Practices (NPP) is available and linked on the site.
[ ] A documented breach notification policy exists.

Mobile Compatibility and Accessibility

[ ] HIPAA protections apply equally to the mobile version of the website.
[ ] The website complies with Web Content Accessibility Guidelines (WCAG 2.1 or higher).

Ongoing Monitoring and Staff Training

[ ] HIPAA compliance and website security are reviewed at least annually.
[ ] Staff with access to website systems are trained in HIPAA rules.
[ ] A plan is in place for responding to security breaches or unauthorized access.

References

1. Hesse BW, Moser RP, Rutten LJ. Surveys of physicians and electronic health information. N Engl J Med. 2010;362(9):859–60.
2. Madanian S, Nakarada-Kordic I, Reay S, Chetty T. Patients' perspectives on digital health tools. PEC Innov. 2023;2:100171. https://doi.org/10.1016/j.pecinn.2023.100171. PMID: 37384154; PMCID: PMC10294099
3. Birnbaum F, Lewis D, Rosen RK, Ranney ML. Patient engagement and the design of digital health. Acad Emerg Med. 2015;22(6):754–6. https://doi.org/10.1111/acem.12692. Epub 2015 May 21. PMID: 25997375; PMCID: PMC4674428
4. Iverson SA, Howard KB, Penney BK. Impact of Internet use on health-related behaviors and the patient-physician relationship: a survey-based study and review. J Am Osteopath Assoc. 2008;108(12):699–711.
5. Tonsaker T, Bartlett G, Trpkov C. Health information on the Internet: gold mine or minefield? Can Fam Physician. 2014;60(5):407–8. PMID: 24828994; PMCID: PMC4020634
6. Manteghinejad A, Javanmard SH. Challenges, and opportunities of digital health in a post-COVID19 world. J Res Med Sci. 2021;26:11. https://doi.org/10.4103/jrms.JRMS_1255_20. PMID: 34084190; PMCID: PMC8103966
7. Mumtaz H, Riaz MH, Wajid H, Saqib M, Zeeshan MH, Khan SE, Chauhan YR, Sohail H, Vohra LI. Current challenges and potential solutions to the use of digital health technologies in evidence generation: a narrative review. Front Digit Health. 2023;5:1203945. https://doi.org/10.3389/fdgth.2023.1203945. PMID: 37840685; PMCID: PMC10568450

Chapter 3
Navigating Compliance and Regulations in Digital Marketing

Understanding HIPAA and Other Healthcare Marketing Regulations

HIPAA, the Health Insurance Portability and Accountability Act of 1996, establishes the benchmark for the protection of sensitive patient data in the United States. It mandates that any entity dealing with protected health information (PHI) must implement comprehensive physical, network, and process security measures to safeguard this information.

Side Bar: What Are PHI and ePHI?
Protected health information (PHI) refers to any medical data that can be linked to an individual, which includes names, addresses, birthdates, test results, diagnoses, insurance numbers, and more. When that data is created, stored, or transmitted electronically (such as via email, a patient portal, or electronic health record), it becomes electronic protected health information (ePHI).

Both PHI and ePHI are protected under the Health Insurance Portability and Accountability Act (HIPAA). Mishandling this information, even unintentionally, can result in significant legal, financial, and reputational harm for a medical practice.

The significance of HIPAA in digital healthcare marketing cannot be overstated, as it directly impacts how patient information is used and shared in marketing efforts.

HIPAA sets the standard for protecting sensitive patient data. Any entity dealing with protected health information (PHI) must ensure that all the required physical, network,

C. Quinn, *The Digital Practice*, https://doi.org/10.1007/978-3-032-07745-5_3

and process security measures are in place and followed.[1] To set the standard, HIPAA lays down critical rules to protect personal health information.[2] Here's a breakdown:

- **Privacy rule**: This rule serves as the model for how personal medical records and health information should be treated. It restricts how this sensitive data can be used and shared while also granting your patients rights over their own information. This includes how their data might be utilized in digital marketing efforts.
- **Security rule**: This focuses on the digital realm, ensuring that any electronic personal health information (ePHI) is kept confidential and secure. It mandates that healthcare providers put in place robust security measures to protect this information from any potential breaches.
- **Breach notification rule**: In the unfortunate event of a data breach involving unprotected health information, this rule outlines the steps that must be taken. Affected individuals, the HHS, and sometimes even the media must be notified, ensuring transparency and accountability.

The importance of HIPAA in digital healthcare marketing lies in its role in maintaining patient trust and confidentiality. By ensuring that all marketing practices are HIPAA-compliant, healthcare providers and marketers can prevent unauthorized access to PHI. This compliance is important not only for legal adherence but also for building and maintaining trust with patients. When patients feel confident that their personal health information is secure, they are more likely to engage with digital healthcare services and platforms.

Moreover, HIPAA compliance in digital marketing helps healthcare entities navigate the complex landscape of digital privacy laws. It provides a clear framework for using patient data in marketing, ensuring that communications are ethical and do not compromise patient privacy. This framework includes obtaining explicit consent from patients before using their information in marketing materials, a practice that reinforces trust and transparency.

These core HIPAA rules form the foundation of patient privacy obligations. In digital marketing, their application becomes more complex, and more critical.

Sidebar: What Does "Documented Consent" Look Like?

Documented consent means that the patient has clearly agreed to a specific use of their protected health information (PHI) and that agreement has been captured in a verifiable, retrievable format. This can include the following:

- **Written consent** on a paper form signed by the patient.
- **Digital consent** is captured through secure patient portals, checkboxes on online forms, or e-signature tools.

(continued)

[1] https://www.digitalauthority.me/resources/healthcare-digital-marketing-and-hipaa/. Accessed 3/5/24.

[2] https://tribalvision.com/navigating-hipaa-compliance-for-digital-marketers/. Accessed 3/5/24.

- **Form-based consent** that is embedded into intake or communication processes, such as registration paperwork or telehealth agreements.

  To be valid under HIPAA, the consent must specify the following:

- What information is being shared
- Who is receiving it
- For what purpose
- An expiration date (when applicable)

The documentation should be stored in the patient's record and accessible for auditing, legal review, or patient inquiry.

Figure 3.1 illustrates a typical workflow for collecting, storing, and referencing patient consent in compliance with HIPAA requirements.

Fig. 3.1 Patient consent workflow. (Image Quinn/ChatGPT)

Patient Privacy and Digital Marketing Compliance

Patient confidentiality remains a foundational obligation in healthcare. This principle extends to digital marketing. Healthcare providers must ensure that no content, message, or campaign discloses protected health information (PHI) without proper authorization. HIPAA defines strict requirements for the use and protection of PHI in both clinical and nonclinical settings.

In the context of marketing, practices must secure documented and specific consent from patients before using any information that could identify them. For example, a provider who wants to share a patient's story or photograph on the practice website or social media channels must obtain clear, written permission in advance. This consent should outline what information will be shared, who will receive it, and for what purpose. When healthcare organizations follow this process, they not only avoid regulatory penalties but also strengthen the trust patients place in their care.

Despite these restrictions, numerous examples exist of health systems that utilize patient stories responsibly. Cleveland Clinic, for instance, publishes patient interviews on its public site, each backed by formal consent. These stories help humanize healthcare while preserving patient privacy.[3] While HIPAA is the primary framework in the United States, healthcare organizations must also consider additional regulations, especially when operating across state lines or international borders.

HIPAA is not the only law that governs digital privacy. Other regulations may apply depending on location and patient population. The Health Information Technology for Economic and Clinical Health (HITECH) Act reinforces HIPAA rules, especially for electronic health records, and increases the penalties for noncompliance. The General Data Protection Regulation (GDPR) protects the data of patients in the European Union. Even if a healthcare provider is based in the United States, GDPR may apply when international patients are involved. The California Consumer Privacy Act (CCPA) gives California residents more control over the personal data that businesses collect, including healthcare organizations. For a detailed side-by-side comparison of HIPAA, HITECH, GDPR, and CCPA, including scope, consent requirements, enforcement mechanisms, and penalties, see Appendix 3.2.

Several tools can help healthcare practices manage consent and privacy. Electronic signature services such as DocuSign and Adobe Sign offer secure ways to collect written permission from patients. For a comparison of HIPAA-compliant digital tools, including their functions, compliance features, and best use cases, see Appendix 3.1.

Broader privacy management systems, including OneTrust and TrustArc, help ensure that digital campaigns remain compliant with HIPAA, HITECH, GDPR, and CCPA. These platforms support documentation, consent tracking,

[3] https://my.clevelandclinic.org/patient-stories. Accessed 7/27/25.

and secure data collection. In addition to working with trusted vendors, healthcare marketers can protect patient identity through another essential method: de-identification.

See Appendix 3.1 for a comparison of privacy tools. See Appendix 3.2 for a comparison of these privacy laws.

Sidebar: What You Need to Know About Third-Party Vendors and HIPAA

If your practice works with third-party vendors, such as web designers, email platforms, IT consultants, or telehealth providers, you are still responsible for ensuring that they handle protected health information (PHI) appropriately. Under HIPAA, these companies are considered *business associates* and are legally required to follow strict privacy and security rules.

Before giving a vendor access to any PHI, you must have a signed *Business Associate Agreement* (BAA) in place. This document outlines the vendor's responsibilities for protecting patient data and complying with HIPAA standards. Without a BAA, you risk severe penalties if a data breach occurs, even if the error was not your fault.

Make sure your vendors use encryption, limit access to PHI, monitor systems for security threats, and report any breaches immediately. You should also verify that they comply with HIPAA's Security Rule, which includes technical safeguards such as access controls and audit logs.

Always confirm that your vendors offer HIPAA-compliant services and are willing to sign a BAA, and if they will not, do not use them. Protecting patient privacy isn't just about technology; it's about choosing the right partners.

Understanding De-identification in Healthcare Marketing

Another important safeguard is de-identification, which refers to removing personal details from patient data so that it cannot be traced back to an individual. This is particularly relevant when using patient stories, images, or health information in marketing materials. HIPAA outlines two approved methods. The Safe Harbor method requires the removal of 18 specific identifiers, including names, birth dates, and contact information. The Expert Determination method relies on a qualified statistician to confirm that the likelihood of identification is very low. It is not enough to delete a name alone. Patients can sometimes be identified through combinations of details such as location, age, or medical condition. For practices that do not obtain written patient permission, proper de-identification helps protect patient privacy and reduces the risk of noncompliance.

Sidebar: Two Methods of De-identifying Patient Data

To use patient data in marketing or research without explicit authorization, HIPAA allows healthcare providers to remove identifiers through one of two approved methods:

1. **Safe Harbor Method**
   This approach requires the removal of 18 specific identifiers from the dataset. These include names, geographic subdivisions smaller than a state (like ZIP codes), all elements of dates related to the individual (such as birth date or admission date), phone numbers, email addresses, Social Security numbers, medical record numbers, and other direct or indirect identifiers. Once these elements are removed, and the provider does not have actual knowledge that the data could be used to identify the patient, the information is considered de-identified.
2. **Expert Determination Method**
   This method involves hiring a qualified statistician or data expert who uses accepted scientific principles and statistical methods to analyze the risk of identification. If the expert determines that the likelihood of someone being identified from the data is very small—and documents the methods and results—the data is considered de-identified. This method allows more flexibility but requires professional expertise and documentation.

Both methods help protect patient privacy while allowing healthcare practices to use data responsibly. However, neither is a substitute for patient consent when identifiable information is involved.

By understanding these privacy obligations and using appropriate safeguards, healthcare professionals can develop effective marketing strategies that respect patient rights. Following these steps helps maintain legal compliance while reinforcing the ethical duty to protect confidentiality. With a clearer understanding of regulatory responsibilities, the next challenge lies in applying these principles to high-risk environments, particularly social media.

Table 3.1 compares the scope, consent requirements, breach notification rules, penalties, and enforcement bodies for HIPAA, HITECH, and GDPR.

For additional comparisons including the California Consumer Privacy Act (CCPA), see Appendix 3.2.

Table 3.1 Comparison of HIPAA, HITECH, and GDPR

Feature	HIPAA	HITECH	GDPR
Full name	Health Insurance Portability and Accountability Act	Health Information Technology for Economic and Clinical Health Act	General Data Protection Regulation
Region of application	United States	United States	European Union (applies globally to data of EU citizens)
Year enacted	1996	2009	2016 (effective 2018)
Primary purpose	Protects the privacy and security of health information	Strengthens HIPAA by promoting secure adoption of electronic records	Regulates the collection and processing of personal data
Focus of regulation	Protected health information (PHI)	Electronic health records (EHRs), breach notification	Personal data, including health, financial, and biometric information
Consent requirements	Requires patient authorization for PHI use beyond treatment or payment	Requires breach notification and tracking of electronic PHI	Requires clear, affirmative consent before collecting personal data
Breach notification	Required within 60 days of discovery	Strengthens notification rules, adds reporting requirements	Required within 72 hours
Penalties for noncompliance	Civil and criminal penalties	Increased penalties and mandatory audits	Fines up to €20 million or 4% of global revenue
Right to access	Patients can request their medical records	Encourages electronic access	Individuals have broad rights to access, correct, or erase their data
Key enforcement bodies	US Department of Health and Human Services (HHS), Office for Civil Rights (OCR)	HHS, Office of the National Coordinator for Health IT (ONC)	Data Protection Authorities in each EU member state
Applies to	Covered entities and business associates	Healthcare providers using EHRs	Any entity processing data of EU residents, including US companies

Social Media

In the realm of digital marketing for healthcare, social media stands as a powerful tool to engage and inform patients. However, the use of these platforms comes with its own set of challenges, particularly in terms of compliance with privacy laws. Healthcare professionals must exercise caution to ensure their online interactions do not compromise patient confidentiality or violate regulations, such as HIPAA.

One common pitfall is the inadvertent disclosure of protected health information (PHI) while sharing success stories or testimonials. To avoid this, always secure explicit consent from patients before sharing any aspect of their care or treatment outcomes online. This consent should be documented and stored securely, reflecting the patient's agreement to the use of their information in your marketing materials. Alternatively, healthcare workers have taken videos of each other while discussing a patient's medical status, using the patient's name.

Another pitfall is posting pictures or videos of patients on social media without their consent. The Internet is awash in instances where healthcare providers have taken photos of patients and posted them online, thinking that they would only be shared among friends. Some examples include the following:

- At a nursing home in New Jersey, a nursing assistant photographed a resident's genitals and sent the picture to a friend, who uploaded it to Facebook.[4]
- A nurse's aide in a nursing home in upstate NY took photos of an incontinent resident's genitals covered with fecal matter and shared them with another staff member on Snapchat.[4]
- A group of resident doctors shared photos taken during surgeries on Instagram, displaying removed body parts without patient consent. Some images, showing patients mid-surgery, raised concerns about potential patient identification, highlighting a clear disregard for HIPAA's privacy rule.[5]
- In Atlanta, Georgia, a dermatological surgeon took videos of herself dancing in the OR while her patient was visible on the operating table and posted them on Instagram and YouTube.[6]

A strict policy limiting the use of recording equipment around healthcare facilities should be enforced.

Best practices for engaging on social media include creating content that is general and informative rather than patient-specific. For instance, instead of posting about a specific patient's recovery journey, healthcare marketers might share general health tips, news about advancements in medical research, or information about the services they offer. This approach helps maintain patient privacy while still providing valuable content to your audience.

Another best practice is to implement a clear social media policy for your practice. This policy should outline what is acceptable to post, who is authorized to post on behalf of your organization, and how to respond to patient inquiries or comments online. It's crucial to train all staff members on these guidelines to ensure consistency and compliance across all digital interactions.

[4] https://www.propublica.org/article/inappropriate-social-media-posts-by-nursing-home-workers-detailed. Accessed 3/5/24.

[5] https://etactics.com/blog/social-media-hipaa-violations

[6] https://www.nytimes.com/2018/06/07/us/video-doctor-surgery-dancing.html

Moreover, employing privacy software and tools can help monitor and manage your social media presence more effectively. These tools can flag potential HIPAA violations before they occur, offering an additional layer of protection.

Real-life examples of successful social media engagement that respects patient privacy include wellness tips shared by healthcare providers, Q&A sessions where general health questions are answered, and awareness campaigns about public health issues. These strategies enable healthcare professionals to harness the power of social media responsibly, promoting patient engagement while adhering to relevant privacy laws. Although social media offers powerful opportunities for engagement, it also introduces specific risks that demand special attention.

Addressing Social Media Challenges

As social media becomes an integral part of digital outreach, physician practices must carefully balance visibility with legal and ethical obligations. Platforms like Facebook, Instagram, TikTok, and X (formerly Twitter) allow healthcare providers to connect with current and potential patients, build brand identity, and share educational content. However, this visibility comes with significant regulatory responsibilities, particularly under HIPAA.

One of the most common social media pitfalls is the inadvertent disclosure of protected health information (PHI). Even when names are omitted, posts that include details about patient conditions, treatments, or interactions can be identifiable, especially in small or rural communities. According to the US Department of Health and Human Services (HHS), a healthcare provider's social media activity must never include any information that could reasonably identify a patient unless explicit, documented consent has been obtained [1].

Another risk involves responding to online reviews or comments. While patients may choose to share their health experiences publicly, healthcare providers must not respond in a way that confirms the individual's status as a patient. A 2019 HIPAA enforcement action resulted in a $10,000 fine against a dental practice for disclosing PHI while responding to a Yelp review, even though the comment was brief and well-intentioned [2].

To remain compliant, physicians and staff should follow key best practices:

- **Avoid discussing individual cases or clinical experiences on social platforms**, even in anonymized form.
- **Refrain from responding to reviews with clinical details or patient acknowledgment.**
- **Use clear disclaimers** on social profiles indicating that the platform is for informational purposes only and not for medical advice.
- **Implement a social media policy** for staff and vendors outlining appropriate use, content approval workflows, and training requirements.

- **Limit access to account management** to individuals trained in HIPAA compliance and digital communication strategies.

Social media management tools such as Sprout Social or Hootsuite can assist in scheduling, moderation, and content approval, but they must be configured in compliance with HIPAA if any patient interactions occur. When posting testimonials or patient stories, ensure that signed, documented consent is stored and retrievable.

Services like Adobe Sign and DocuSign offer HIPAA-compliant e-signature platforms that are suitable for these purposes [3].

Finally, healthcare practices should regularly audit their social media presence for risks. This includes reviewing prior posts, monitoring comments for inadvertent PHI disclosures, and revising content that may no longer meet privacy standards. Periodic training and policy reviews can reduce the likelihood of violations while empowering practices to maintain a trusted and informative online presence.

By taking a proactive and educated approach, physician practices can successfully navigate the challenges of social media while protecting patient privacy and maintaining legal compliance. To help practices implement these strategies effectively, a range of digital tools can translate compliance into action.

Interactive Tools: Making Compliance Actionable for Your Practice

As a physician or practice manager expanding into digital marketing, you may already be aware of HIPAA and other privacy regulations, but ensuring compliance across websites, emails, and social media can feel overwhelming. Interactive tools such as checklists, self-assessments, and online quizzes can help translate these legal requirements into clear, actionable steps for your office.

For example, a HIPAA compliance checklist can guide you through key questions, such as the following: Does your website use encryption? Do you have patient consent documented for testimonials? Are staff members trained on the appropriate use of practice social media accounts? These tools serve as quick diagnostics to identify potential gaps in your current workflow.

Here's a sample compliance self-check tool your practice can use internally or during staff meetings.

Table 3.2 presents a quick HIPAA compliance self-check for medical marketing, which practices can use internally to identify privacy risks.

Interactive quizzes can also help train staff on critical marketing topics, such as when consent is required, how data tracking tools may impact privacy, or how to respond to patient reviews online. These formats make it easier to educate your team without long policy manuals or time-consuming training sessions.

Many reputable sources, including professional associations, legal advisors, and healthcare marketing platforms, offer downloadable or embeddable compliance tools that can be customized to suit your practice. They not only enhance your

Table 3.2 Quick HIPAA compliance self-check for medical marketing

	Question	Yes	No
1.	Do we have documented patient consent for all testimonials used online?		
2.	Are our social media posts reviewed for possible PHI before publishing?		
3.	Have we signed a Business Associate Agreement (BAA) with all third-party vendors (e.g., email, website, CRM)?		
4.	Is our website using HIPAA-compliant encryption and secure data storage?		
5.	Are all staff trained on appropriate digital communication and marketing practices?		
6.	Do we have a clear policy for managing patient reviews and responses on public platforms?		

Instructions: Use this six-question checklist to spot potential privacy risks in your current marketing strategy. For each item, check YES or NO
Scoring
•**5–6 YES**: You're on the right track. Continue reviewing policies regularly
•**3–4 YES**: Moderate risk. Review training, consent procedures, and vendor contracts
•**0–2 YES**: High risk. Consider a full marketing compliance audit

ability to identify risks but also signal to patients that your practice prioritizes privacy and transparency. See Appendix 3.1 for a comparison of platforms commonly used to support HIPAA-compliant consent and data workflows.

For an overview of HIPAA-compliant digital tools, see Appendix 3.1, which compares functions, compliance features, and best use cases for common platforms.

Rather than approaching compliance as a static checklist or abstract concept, interactive content encourages active engagement. It helps you and your team take ownership of privacy practices while making the learning process more efficient and aligned with your marketing goals. These tools are especially useful when launching new campaigns or updating your digital presence. Even with strong policies and tools in place, staying compliant requires continuous awareness of regulatory updates.

Staying Up to Date with Healthcare Marketing Regulations

Healthcare marketing rules change frequently. For physicians and practice managers, staying informed is important for legal compliance and for protecting patient relationships.

HIPAA and HITECH are the primary federal laws covering patient privacy and data security. These laws are sometimes updated to reflect changes in how practices use electronic health records or communicate with patients online. In addition to federal rules, several states have created their own privacy laws. California's Consumer Privacy Act (CCPA), for example, requires businesses to give patients more control over how their personal data is used.

If your practice serves international patients or operates a telehealth program across borders, you may also need to follow the General Data Protection Regulation

(GDPR). This European Union law requires organizations to protect the personal data of EU citizens, no matter where the organization is based.

To keep up with changes, consider subscribing to official resources. The Office for Civil Rights at the US Department of Health and Human Services provides updates on HIPAA.[7] Other reliable sources include HealthIT.gov and the American Medical Association's digital policy page. Legal firms like Foley & Lardner[8] or BakerHostetler[9] also publish blogs and newsletters written for healthcare professionals.

When working with vendors, such as web developers, email platforms, or CRMs, ask whether their systems are compliant with HIPAA and state laws. If they offer alerts when regulations change, make sure those updates are forwarded to your office.

For example, in 2023, the Federal Trade Commission clarified that using tracking pixels from platforms like Facebook or Google on patient-facing web pages could count as a privacy violation. Practices that used these tools without consent faced fines and reputational harm.

You do not need legal training to stay compliant. Regular reading and careful selection of vendors can help keep your practice in line with current rules. Staying informed shows patients that your team takes their privacy seriously and reduces the risk of costly errors. Beyond legal requirements, ethical considerations shape how digital outreach is perceived by patients and the public.

Best Practices for Ethical Marketing and Maintaining Patient Privacy

Ethical digital marketing in healthcare begins with securing informed consent from patients before using any identifiable health information for promotional purposes. This step not only meets HIPAA's legal requirements but also reinforces a practice's commitment to respecting patient autonomy and privacy. Consent should be clearly documented and should include what information is being shared, with whom, and for what purpose.[10]

---

[7] hhs.gov/hipaa. Accessed 7/27/25.

[8] https://www.foley.com/practice-areas/corporate/health-care/health-care-compliance-and-operations/health-privacy-and-security/. Accessed 7/27/25.

[9] https://www.healthlawupdate.com/. Accessed 7/27/25.

[10] https://www.knbcomm.com/blog/ethical-considerations-in-healthcare-marketing-best-practices-for-achieving-roi-responsibly

To protect patient data, healthcare practices should consider using digital privacy tools such as OneTrust[11] or TrustArc,[12] which help monitor and manage data collection practices in line with HIPAA, HITECH, and GDPR regulations. When obtaining electronic consent, platforms like Adobe Sign or DocuSign can ensure secure transmission and audit trails. For organizations handling large volumes of sensitive data or operating across state lines, employing a data privacy officer may provide helpful oversight and ensure policies remain current with evolving standards.[13]

Marketing materials should avoid exaggeration and medical claims that lack evidence-based support. Communications should reflect the tone and standards found in clinical care, clear, respectful, and grounded in fact. The American Medical Association's Code of Medical Ethics advises that all advertising must be truthful and not misleading, especially when referencing outcomes or treatments.[14]

Ongoing training is essential to maintain compliance.[15] Practices can offer quarterly updates or subscribe to healthcare legal newsletters to stay informed about regulatory changes. In smaller practices, a designated staff member can track updates from trusted sources such as the Department of Health and Human Services (HHS.gov), the AMA, or health law blogs from firms like Foley & Lardner or BakerHostetler, which regularly publish compliance guidance for clinicians and healthcare organizations.

By adhering to these guidelines, healthcare providers and marketers can navigate the complex regulatory environment of digital marketing, ensuring that their practices are not only effective but also compliant and ethical. This approach safeguards patient privacy, upholds the integrity of the healthcare profession, and builds lasting trust with the audience. Ultimately, rather than undermining patient trust, combining regulatory compliance with ethical awareness ensures that marketing efforts support it.

---

[11] https://www.onetrust.com/solutions/privacy-automation/?ef_id=Cj0KCQjwnJfEBhCzARIsAIMtfKKsgVflZPTRYtPfM_1g8zgQf-24lmWkGxklGosbIPgdtECnJO_e83YaAvNjEALw_wcB:G:s&s_kwcid=AL!17820!3!726716293337!e!!g!!onetrust%20privacy!22059060923!176186793481&utm_source=google&utm_medium=cpc&utm_campaign=G|NA|Search|Brand|PrivacyAutomation|US&utm_content=Solutions_PrivacyAutomation&utm_term=onetrust%20privacy&gad_source=1&gad_campaignid=22059060923&gbraid=0AAAAAobCVrLPnsm_SNyoEIm1CeoVZmlHG&gclid=Cj0KCQjwnJfEBhCzARIsAIMtfKKsgVflZPTRYtPfM_1g8zgQf-24lmWkGxklGosbIPgdtECnJO_e83YaAvNjEALw_wcB. Accessed 7/27/25.

[12] https://trustarc.com/. Accessed 7/27/25.

[13] https://www.ed.gov/about/ed-offices/ocr. Accessed 7/27/25.

[14] https://code-medical-ethics.ama-assn.org/. Accessed 7/27/25.

[15] https://sproutsocial.com/insights/guides/social-media-compliance/. Accessed 7/27/25.

Conclusion

Physicians and healthcare marketers operate in a digital environment that offers both unprecedented reach and heightened responsibility. By understanding the regulatory landscape, documenting patient consent, choosing the right tools and vendors, and adopting ethical communication practices, your practice can maintain compliance while building trust with patients. Compliance is not a one-time task but an ongoing commitment that aligns marketing strategy with professional integrity.

Appendixes

Appendix 3.1: Comparison of HIPAA-Compliant Digital Tools

Tool	Primary function	HIPAA compliance	BAA availability	Key features	Best use cases
DocuSign	Electronic signatures	Yes, with signed BAA	Available for healthcare providers	Signer authentication, audit trails, mobile access	Patient consent forms, intake paperwork, BAAs
Adobe Sign	Electronic signatures	Yes, with signed BAA	Available for covered entities	Microsoft 365 integration, mobile workflows, tracking and alerts	Consent documentation, EHR integration, contracts
OneTrust	Privacy and compliance management	Yes, with enterprise configuration	Offered for enterprise clients	Consent management, data mapping, cookie banners, incident tracking	Enterprise-level compliance management
TrustArc	Data governance and privacy compliance	Yes, with enterprise configuration	Offered for enterprise clients	Risk assessments, privacy reporting, vendor risk monitoring, DSAR[a] management	Large practices with complex data workflows

Note: All four platforms support HIPAA compliance when configured properly and accompanied by a signed Business Associate Agreement (BAA). Practices should confirm the compliance status and security protocols of any third-party service before onboarding

[a]A **Data Subject Access Request (DSAR)** is a formal request made by an individual to access the personal information that an organization holds about them. Under privacy regulations like the **General Data Protection Regulation (GDPR)** and **California Consumer Privacy Act (CCPA)**, individuals have the right to request access to their data, ask for corrections, or request deletion

Appendix: 3.2:Global Privacy Laws Comparison Table

Feature	HIPAA (USA)	HITECH (USA)	GDPR (EU)	CCPA (California)
Full name	Health Insurance Portability and Accountability Act	Health Information Technology for Economic and Clinical Health Act	General Data Protection Regulation	California Consumer Privacy Act
Effective date	1996	2009	May 25, 2018	January 1, 2020
Scope	Healthcare providers, plans, and clearinghouses	Extends HIPAA, applies to EHR systems and data breaches	All organizations processing personal data of EU residents	For-profit businesses handling California residents' data
Focus	Protected health information (PHI)	Electronic Health Records (EHRs) & breach notification	Broad personal data protection and privacy	Consumer data privacy and control
Consent requirement	Implied for treatment; written for marketing or sharing	Reinforces HIPAA standards	Explicit, informed consent for data collection and processing	Opt-out model; must allow consumer control over data use
Right to access data	Yes, within 30 days	Yes, especially EHR-related	Yes, comprehensive rights including access and portability	Yes, consumers may request access to collected data
Right to be forgotten	No	No	Yes	Limited—mostly opt-out of sale
Breach notification	Within 60 days to HHS and affected individuals	Mandatory; includes financial penalties	Within 72 hours to supervisory authority and subjects	Reasonable notice required
Penalties for violations	Civil and criminal penalties	Higher penalties than HIPAA alone	Up to €20 million or 4% of global annual revenue	Fines up to $7,500 per violation
Business associate rules	Required contracts (BAAs)	Same as HIPAA	Data processors must have binding contracts	Service providers must comply if data is sold or shared
Data subject rights	Access, amendments, accounting of disclosures	Same as HIPAA	Access, rectification, erasure, restriction, objection, portability	Access, deletion, and opt-out of sale

References

1. U.S. Department of Health and Human Services (2023) HIPAA and social media. https://www.hhs.gov/hipaa/for-professionals/privacy/guidance/hipaa-social-media/index.html
2. Office for Civil Rights (2019) OCR settles first HIPAA case involving a patient complaint about a provider's response to a Yelp review. https://www.hhs.gov/about/news/2019/10/02/ocr-settles-first-hipaa-case-involving-yelp-review.html
3. Adobe (2023) Adobe sign and HIPAA readiness. https://helpx.adobe.com/sign/kb/hipaa-compliance.html

Chapter 4
Integrating Digital and Traditional Marketing Strategies

Coordinated marketing creates a unified strategy, promoting a brand consistently across various platforms with a coherent message and shared goal. It's the key to forging a strong brand image and message, essential for cultivating brand recognition and trust. By aligning all marketing efforts, businesses not only sharpen their competitive edge but also boost the efficiency, consistency, and measurability of their marketing activities.

In the realm of traditional marketing, methods such as print ads, television commercials, and direct mail have long been the dominant ones. Yet, as digital marketing gains traction, these conventional strategies are gradually losing their appeal. Digital marketing, through its use of online platforms such as websites, social media, and email, offers a distinct advantage. It enables the precise targeting of potential customers, significantly enhancing reach and efficiency with minimal effort.

Sticking to just one marketing channel, however, can stifle a brand's potential. Today, consumers engage with brands through a myriad of platforms and touchpoints. A unified marketing strategy that weaves together both traditional and digital efforts can more effectively captivate the target audience, wherever they may be.

Such a blended approach not only ensures a better return on investment by leveraging the strengths of each marketing channel but also ensures that messages resonate with consumers across all platforms. Synchronized marketing amplifies brand visibility and engagement, delivering a consistent message that fosters brand loyalty and boosts conversion rates. It smartly navigates the complex landscape of consumer behavior, opting for a versatile presence over confinement to a single marketing avenue.

The following sections offer practical combinations of digital and traditional tactics that medical practices can adopt today. The key to success lies not in choosing one format over another, but in thoughtfully combining both. The next section outlines how to do that with specific, real-world strategies

C. Quinn, *The Digital Practice*, https://doi.org/10.1007/978-3-032-07745-5_4

Strategies for a Cohesive Marketing Approach

Integrating digital and traditional marketing strategies is essential for medical practices aiming to maximize their reach and impact. One practical approach is to leverage the combination of billboards and geotargeting. This strategy utilizes the broad visibility of billboards along with the precision of digital geotargeting to engage both wide and specific audiences effectively.

Leverage Billboards and Geotargeting

Combining traditional billboards with digital geotargeting enables medical practices to reach local audiences more efficiently and track engagement in ways previously impossible with print alone. This approach fosters broad community awareness while targeting specific geographic areas, a distinct advantage for practices serving local populations.

Billboards have been a staple in traditional advertising because of their significant reach and visibility. They are strategically placed in high-traffic areas, ensuring that a large number of people see the advertisement. However, the main limitation of billboards is their lack of targeting; they are static and cannot change based on the viewer's demographics or interests.

Enter geotargeting. Geotargeting is a digital marketing strategy that uses a person's geographic location, often determined by GPS, IP address, or mobile device data, to deliver tailored content, advertisements, or promotions. In healthcare marketing, geotargeting enables practices to reach potential patients within a specific radius, zip code, or neighborhood, improving the relevance and effectiveness of outreach. For example, a dermatology clinic in Boston might run a Facebook ad promoting same-day appointments but only show it to users within a 10-mile radius of the office. This approach reduces wasted ad spend and increases the likelihood that the message reaches people who are both interested and geographically able to respond.

Figure 4.1 illustrates how traditional billboard exposure can be enhanced with geotargeted digital ads to deliver more personalized outreach to nearby audiences.

By integrating geotargeting with billboard advertising, medical practices can enhance the effectiveness of their marketing efforts. For example, a billboard can promote a general message about the medical practice. At the same time, geotargeted ads on smartphones and other digital devices can provide tailored messages to individuals within the vicinity of the billboard. This dual approach ensures that while the billboard captures the attention of a broad audience, the geotargeted ads refine the message for specific demographic groups, increasing the relevance and potential impact of the advertisement.

Consider a medical practice specializing in family healthcare. A billboard near a busy intersection advertises the practice's commitment to comprehensive family

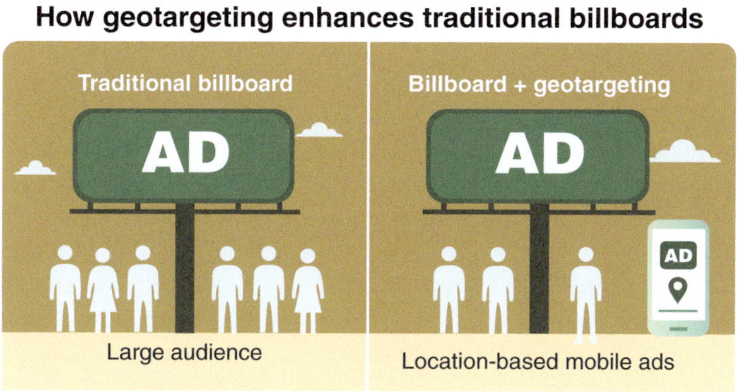

Fig. 4.1 Integrated billboard and geotargeting workflow for medical practices. (Image by Quinn/ChatGPT)

health services. Simultaneously, parents within a certain radius of the billboard receive geotargeted ads on their mobile devices highlighting pediatric care services offered by the practice. This targeted approach ensures that the message reaches the intended audience, increasing the likelihood of engagement.

By integrating traditional and digital marketing strategies, such as billboards and geotargeting, medical practices can leverage the strengths of both approaches. This cross-platform marketing strategy ensures broad visibility while delivering targeted, relevant messages to specific audience segments, thereby maximizing both reach and impact. While billboards and geotargeting are ideal for high-visibility, location-based outreach, print media offers a different strength, depth of message, and physical presence. Let's look at how magazines and digital content can work together to extend reach and engagement.

Magazines and Digital Content

Successful campaigns effectively combine print and digital elements to extend reach and reinforce messaging across multiple platforms. By leveraging magazine advertisements, medical practices can harness the extensive reach of traditional media to introduce their services to a broader audience. This strategy not only builds brand awareness but also fosters a tangible connection with potential clients.

Enhancing this approach, digital content can complement magazine ads by incorporating QR codes or social media links, facilitating a seamless transition from print to digital platforms.

A QR code, or Quick Response code, is a square barcode that connects users to digital content when scanned with a smartphone camera. It directs them to a website, online form, video, or other resource without needing to type in a web address.

Healthcare practices use QR codes on printed materials such as brochures, appointment cards, posters, or magazine ads. This lets patients act immediately, whether by booking an appointment, watching an informational video, or reading about a service.

QR codes also support tracking. Each code can include built-in tags that collect data on how often it was scanned, how long visitors stayed on the linked page, what actions they took, and what kind of device they used. Practices can use this information to see which materials are effective and adjust their marketing based on real results.

Furthermore, embedding a clear call to action in both digital and traditional campaigns can significantly improve engagement tracking and effectiveness. Lastly, targeting magazine content to align with the specific interests of the intended audience, coupled with personalized digital marketing strategies, can increase engagement and improve campaign effectiveness. This holistic approach ensures a comprehensive marketing strategy that leverages the strengths of both traditional and digital media, providing a well-rounded approach.

To complement this, enhancing your online presence with detailed digital content allows for deeper engagement. This content can range from informative blog posts and detailed service descriptions to patient testimonials and educational videos. Such digital materials serve to educate further and engage potential patients, drawing them closer to your practice.

Content creation and maintenance can be handled in different ways depending on the size and resources of a medical practice. Some practices rely on in-house staff, such as a marketing coordinator or communications officer, to draft articles, update website copy, or manage digital newsletters. Others contract with third-party marketing firms that specialize in healthcare content.

These vendors often provide services such as keyword research, content calendars, search engine optimization, and performance tracking. Regardless of who produces the material, the practice should maintain editorial oversight to ensure clinical accuracy, legal compliance, and brand alignment.

This dual approach not only broadens your reach but also enhances the consumer journey. Starting with the awareness sparked by a magazine ad, potential patients are then drawn into a more engaged relationship with your practice through your digital content. This journey, from initial awareness to decision-making, is enriched by the narrative that unfolds across these platforms, positioning your practice as a trusted authority in your field.

In essence, leveraging both magazine ads and digital content enables medical practices to create a more dynamic and effective marketing strategy. This consolidated approach ensures that your message remains consistent across various media, thereby enhancing brand recognition and trust among your target audience. Magazines help you target specific interests, but newspapers still carry influence among broad, local audiences. When paired with digital advertising, newspapers can help bridge generational and media-consumption gaps.

Each media channel offers distinct advantages and trade-offs. Table 4.1 summarizes strengths, limitations, and best uses for common channels in healthcare marketing.

Sidebar: Choosing the Right Channel—Strengths and Use Cases in Healthcare Marketing

Table 4.1 Comparative strengths and limitations of traditional and digital marketing channels in healthcare, with recommended use cases for each

Channel	Strength	Limitation	Best used for
Billboards	High local visibility; reinforces name recognition	Static, cannot target specific demographics	General awareness, urgent care, clinic openings
Magazines	Target niche audiences by specialty or interest	High production cost and long lead times	Elective services, private-pay specialties, brand storytelling
Newspapers	Trusted in local and rural communities	Aging readership, limited digital crossover	Physician announcements, Medicare plan info, flu shot promotions
Digital ads (display, SEM)	Precision targeting, measurable ROI, retargeting capabilities	Requires constant monitoring; may be blocked by ad filters	Service line promotion, appointment campaigns, local awareness
Social media	Builds community, enables educational storytelling	Risk of privacy violations; needs staff moderation	Health tips, physician Q&As, reputation building
Email marketing	Personalized outreach, high ROI, easy to automate	May require list building and opt-ins	Recall reminders, newsletter updates, lab result instructions

Note: Ensure all digital platforms are configured for HIPAA compliance when personal health information (PHI) is involved

Newspapers and Online Ads

Magazines offer targeted outreach to niche audiences, while newspapers expand their reach to broader local or regional readers.

Combining newspaper advertising with online ads creates a powerful interaction that enhances the visibility and effectiveness of your marketing efforts. This strategy leverages the credibility and local reach of newspapers, combined with the vast and targeted capabilities of online advertising. By doing so, medical practices can effectively communicate their message to a broader audience while ensuring that it resonates with specific demographics.[1]

Sidebar: Real-World Example—Integrating Print and Digital Ads in Healthcare

CentraCare Health—"Here for Your Health" Campaign

CentraCare, a regional health system in Minnesota, launched the "Here for Your Health" campaign to boost awareness of its preventive services. The campaign targeted both older adults and tech-savvy families through a dual-channel approach:

- **Newspaper advertising:** Print ads ran in local newspapers across Central Minnesota, appealing to older residents and rural populations who rely more heavily on traditional media. These ads featured consistent branding and messaging aligned with digital channels.
- **Online advertising:** CentraCare simultaneously deployed targeted display ads via the Google Display Network and Facebook. These digital ads focused on geofencing and interest-based segmentation to drive traffic to appointment scheduling pages and service-line information.

Key Outcomes

- The campaign generated **over 1 million digital impressions** and led to measurable increases in online appointment requests for preventive screenings.
- Newspaper ads improved brand trust among older demographics, especially those unfamiliar with the health system's digital tools.
- The unified messaging strategy created **cross-channel reinforcement**, helping CentraCare maintain high visibility across media platforms.

This example underscores the value of combining traditional and digital marketing to reach diverse patient populations with consistent, actionable messaging.

Newspaper ads, with their established presence in communities, convey a sense of trustworthiness and reliability. They are particularly effective in reaching an older demographic that still prefers print media over digital platforms. This medium allows for detailed explanations of services, special offers, and the introduction of medical staff, which can significantly impact brand recognition and patient trust.

Online ads, on the other hand, provide the ability to target specific audiences based on demographics, interests, and behavior. They also offer the flexibility of real-time adjustments to optimize campaign performance. Through search engine marketing, social media ads, and display advertising on relevant websites, medical practices can engage with potential patients at various stages of their online journey.

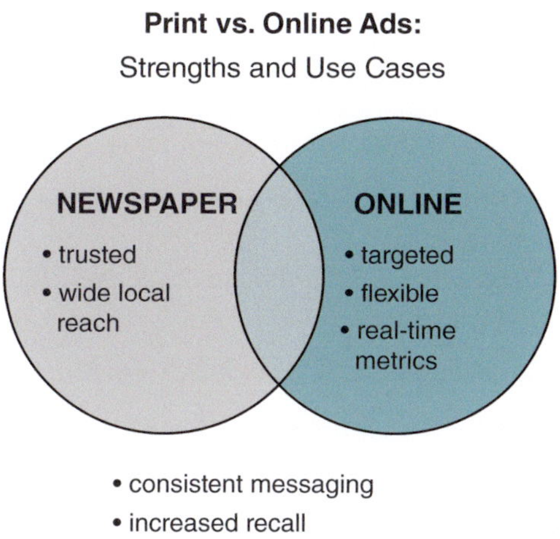

Fig. 4.2 Comparison of print and online advertising strengths in healthcare marketing. (Image by Quinn/ChatGPT)

Figure 4.2 compares the unique benefits of print versus online advertising, highlighting how combining both channels improves visibility and targeting.

By integrating these two approaches, medical practices can ensure that their message is consistently presented across different platforms, reinforcing their brand and increasing the likelihood of recall. This strategy not only maximizes the exposure of their advertising efforts but also supports a cohesive brand narrative that can be crucial for patient decision-making.

By careful planning and execution, the combination of newspapers and online ads can significantly enhance the marketing strategy of any medical practice, leading to increased patient engagement and higher conversion rates. These platform-specific strategies show how integration works in theory. To see how these tactics perform in real-world scenarios, the next section highlights two case studies from healthcare organizations that successfully combined traditional and digital marketing.

As you plan integrated marketing efforts, use the checklist below to ensure your print and digital campaigns align. This tool helps identify gaps in strategy, coordination, and compliance, so your outreach is consistent, effective, and legally sound.

Callout Box: Checklist for Integrating Traditional and Digital Marketing
Use this checklist to assess whether your campaigns are effectively combining offline and online channels.

☐ Print-to-Digital Links

- Have you included a QR code or shortened URL in your print ads?
- Do all printed materials direct users to a landing page or digital resource?

☐ **Consistent Branding**

- Are visual elements (logos, colors, fonts) consistent across print and digital?
- Is your messaging aligned across platforms?

☐ **Cross-Channel Tracking**

- Do you track how many patients engage with digital tools after seeing a print ad?
- Have you set up unique tracking links, UTM codes, or phone numbers for offline promotions?

☐ **Staff, Vendor, and Workflow Coordination**

- Are your marketing vendors aligned on both print and digital executions?
- Is your internal team trained to support both online and offline campaigns?

☐ **Unified Campaign Goals**

- Do both channels drive toward the same patient action (e.g., appointment booking)?
- Are you using analytics to compare performance across media types?

☐ **Channel Relevance and Audience Fit**

- Are you selecting media platforms that match your patients' behaviors and preferences?
- Have you tailored content based on local, generational, or specialty-specific needs?

☐ **Regulatory Compliance**

- Are both print and digital materials reviewed for HIPAA compliance and truth-in-advertising standards?
- Have consent and privacy requirements been addressed where patient data is collected?

The following case studies highlight real-world examples of how integrated strategies drive outcomes in healthcare settings.

Case Studies on Successful Integration Efforts

Case Study 1: Leveraging Social Media to Amplify Medical Practice Visibility

Background
The New Jersey Psychiatric Association (NJPA), a pivotal organization dedicated to the psychiatric profession in Neptune, NJ, embarked on a transformative journey to

establish and enhance its social media presence.[1] With a comprehensive mission to support its members and advocate for patient care, NJPA recognized the need to extend its reach and influence through social media platforms.

Challenge

Initially, NJPA faced the significant challenge of having no social media footprint. The association aimed to cultivate a robust social presence on major platforms such as Facebook, LinkedIn, and Instagram to elevate brand awareness and engage more effectively with both professionals and the public.

Strategy Implementation

Paradigm, a marketing firm, was enlisted to spearhead NJPA's social media strategy. The process commenced with thorough market research, including surveys to gauge current members' platform preferences and content interests. This foundational insight guided the selection of social platforms and the creation of a tailored social media calendar, aligning with NJPA's ongoing marketing initiatives.

Content creation was meticulously planned to reflect NJPA's core messages and resonate with its target audience. The collaboration between marketing strategists and the design team produced engaging social media graphics and posts, which were strategically scheduled using Sprout Social. The use of relevant hashtags and engagement with strategic partners further propelled NJPA's visibility and follower growth.

Outcomes

The concerted efforts yielded remarkable results within the first 6 months of 2023:

- Facebook followers surged by 49%.
- Instagram followers increased by 26.9%.
- LinkedIn followers grew by 9.1%.

As shown in Fig. 4.3, NJPA's integrated campaign significantly increased social media engagement across all platforms in just 6 months.

Fig. 4.3 Growth in NJPA's social media reach following campaign launch. (Image by Quinn/ChatGPT)

[1] https://www.paradigmmarketinganddesign.com/psychiatric-association-increases-web-traffic-by-93-44-from-social-media-efforts/

These platforms saw engagement rates far surpassing the average for nonprofits, with Instagram at 9.7%, Facebook at 5.8%, and LinkedIn at 8.7%. Additionally, the social media campaign for the NJ Pediatric Psychiatry Collaborative significantly boosted social impressions by 339.74%, and website traffic from social media sources soared by 93.44%.

Conclusion

NJPA's strategic foray into social media not only expanded its digital footprint but also significantly enhanced brand awareness and advocacy efforts. This case study exemplifies the power of a well-orchestrated social media strategy in amplifying the visibility and impact of medical practices. Through targeted content, strategic engagement, and continuous performance analysis, medical practices can leverage social media to foster community, support professional development, and advocate for patient care, mirroring NJPA's success.

What can medical practices learn from this campaign? Here are some actionable takeaways:

Lessons for Your Practice

- **Know your audience:** NJPA focused its message on family mental health, which resonated with the concerns of its target demographic.
- **Invest in integrated campaigns:** Combining print, digital, and transit media created a reinforcing cycle of exposure that built trust and recognition.
- **Leverage storytelling:** Highlighting real-life challenges and solutions helped humanize the brand and improve engagement.
- **Measure response across platforms:** Using QR codes and landing pages allowed NJPA to track interest and adjust messaging in real time.
- **Use geographic targeting:** Placing ads in local transit zones and regional publications helped NJPA connect with high-intent, nearby audiences.

While NJPA's example shows how social media can build brand awareness from scratch, large health systems like Northwell offer a broader look at integration during high-stakes campaigns, such as a full rebrand.

Case Study 2: Northwell Health—Building Brand Awareness Through Integrated Campaigns

Background

Northwell Health, New York's largest healthcare provider, rebranded in 2015 from "North Shore-LIJ Health System." The name change was necessary; the previous name proved confusing and failed to resonate beyond Long Island. Before the rebrand, the organization faced low public recognition despite operating 21 hospitals and numerous outpatient facilities across the region.[2, 3]

Strategy

Northwell launched a multimillion-dollar integrated marketing campaign to establish its new brand identity and connect with New York patients. The campaign included the following:

- Broadcast spots and emotional brand-driven TV and radio ads announcing the new name
- Print and transit ads across New York City and Long Island to reinforce visibility
- Digital outreach, including geotargeted search and display ads, and social media storytelling aligned with broadcast content
- A unified narrative built around real patient stories, innovation, and a renewed brand promise under the tagline *Raising Health*.[4]

Outcomes

- Within 6 months of launch, Northwell's campaign logged over **390 million digital ad impressions**, reaching about 13 million online viewers.
- Online appointment scheduling rose by approximately **80%**.
- Brand awareness reached new heights; patient trust and Likelihood-to-Recommend scores climbed into the **high 90s**, and net promoter scores landed in the **45–47%** range.[5]

Figure 4.4 presents a visual breakdown of the media mix Northwell Health used to support its brand transition and patient outreach.

What can medical practices learn from this campaign? Here are some actionable takeaways.

Lessons for Your Practice

- Align your message across all platforms—print, broadcast, online—to reinforce recognition and build trust.

[2] https://www.monigle.com/blog/northwell-health-rebranding-a-health-care-giant/?utm_source=chatgpt.com. Accessed 7/27/25.

[3] https://askwonder.com/research/health-system-branding-case-studies-jkn372yjs?utm_source=chatgpt.com. Accessed 7/27/25.

[4] https://strategichcmarketing.com/northwell-health-inspires-brand-love-with-resolutions-campaign/?utm_source=chatgpt.com. Accessed 7/27/25.

[5] https://www.mdgsolutions.com/learn-about-multi-location-marketing/how-an-advertising-campaign-revived-a-new-york-hospitals-brand/?utm_source=chatgpt.com. Accessed 7/2725.

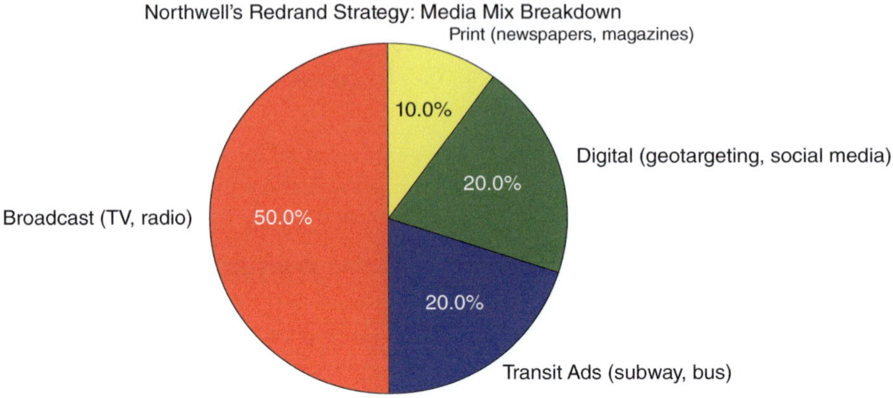

Fig. 4.4 Breakdown of media channels used in Northwell Health's rebranding campaign. (Image by Quinn/ChatGPT)

- Use emotionally resonant storytelling in digital content to complement traditional ads and drive engagement.
- Track performance metrics like online appointment bookings and digital impressions to assess campaign effectiveness.
- Ensure marketing strategy supports broader goals such as rebranding, patient acquisition, or service visibility.

Both case studies reinforce the core idea: marketing channels are more powerful when they work together. Whether you're introducing a new service or redefining your brand, unified campaigns deliver more consistent, memorable results.

Conclusion

Integrated marketing strategies help healthcare practices connect with patients across multiple touchpoints, both online and offline. From billboards and digital ads to print placements and QR-enabled content, the chapter has shown how each channel plays a distinct role in outreach. Case studies demonstrate that when practices align messaging, tailor content to the audience, and measure campaign results, they can significantly enhance visibility, engagement, and trust.

The key is to view marketing as a coordinated effort, not a series of disconnected tactics. By understanding the strengths of each platform and adapting campaigns to local needs, medical practices can enhance their reputation, attract new patients, and support long-term growth. As the next chapter explores digital patient journeys in greater detail, keep in mind how offline and online strategies work best when aligned to patient expectations and behaviors.

Chapter 5
Mapping the Patient Journey in the Digital Age

Introduction

Patients no longer begin their healthcare journey in a waiting room. Today, the process often starts with an online search or a mobile tap. As consumer expectations shift, digital touchpoints now influence how people select providers, schedule appointments, and manage ongoing care.

This chapter examines how patients engage with digital platforms throughout their healthcare journey. It also offers strategies for healthcare professionals to meet patients where they are, on search engines, apps, social media, and telehealth portals. From the first click to follow-up care, every step of the digital patient journey presents an opportunity to improve engagement and build trust.

Understanding how patients begin their healthcare experience online is the first step in adapting your practice to a digital-first environment. To explore this shift more deeply, the next section outlines how digital tools are transforming the way patients engage with healthcare from the start.

Figure 5.1 illustrates the most common digital touchpoints that patients encounter throughout their care journey, from initial symptom searches to follow-up engagement.

C. Quinn, *The Digital Practice*, https://doi.org/10.1007/978-3-032-07745-5_5

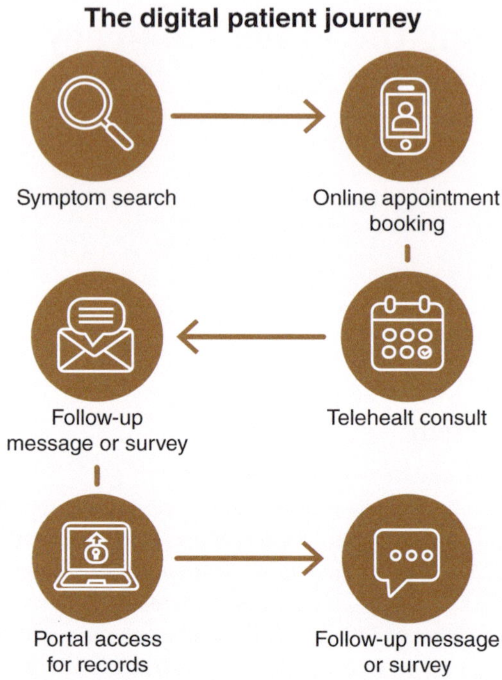

Fig. 5.1 Key digital touchpoints in the modern patient journey. **(Image by Quinn/ChatGPT)**

Digitalization and the Patient Journey

Digital tools are transforming the way patients access and experience healthcare. Most patients begin by searching symptoms online. They use clinic websites, mobile apps, and telehealth services to schedule appointments and manage their records. This process, known as the digital patient journey, facilitates easier access to care and more personalized interactions with healthcare services [1, 2]

Clinics now rely on digital platforms, including websites, social media, apps, and telemedicine, to meet the needs of a connected patient population. These platforms reduce barriers to entry, support better communication, and help patients make more informed decisions [3, 4]. They also give providers insight into how patients engage with the system, which helps build trust and encourages loyalty [5].

Health systems utilize digital strategies to create smoother, more patient-centered experiences. By tracking digital touchpoints, they can identify problems, pinpoint areas for improvement, and enhance both access and follow-through. These improvements support better care and increase efficiency for the practice [6]. Understanding the general impact of digital transformation is only part of the picture. To guide patients effectively, practices must recognize the specific digital touchpoints that shape patient decisions and actions along the care continuum.

Identifying Key Digital Touchpoints

Patient interaction with healthcare services now involves a series of digital steps. Identifying the most critical touchpoints, such as websites, social channels, telehealth platforms, and mobile apps, helps providers understand and shape the patient journey. A 2023 study by Davies et al. emphasized the importance of mapping these interactions to understand how patients navigate the healthcare system. The study supported efforts to build patient-centered care models and helped providers close service gaps more effectively [7].

Each digital platform presents a unique opportunity to reach patients. Symptom searches often lead to clinic websites. From there, patients may schedule appointments, join video visits, and access their records. Providers who pay attention to these moments of contact can respond more effectively to patient needs [8].

Clinics now study behavior patterns, track feedback, and monitor referral traffic. They use this data to guide content and design choices. For example, one dermatology group redesigned its website for mobile use after noticing a rise in smartphone traffic. The update made it easier for users to find information and book visits. Patient portals offer another key access point. These secure platforms let patients view their records, request refills, and read educational content [9]. They also support HIPAA-compliant messaging and allow for asynchronous care when appropriate [10].

By understanding how patients navigate this digital journey, healthcare providers can enhance engagement, strengthen communication, and minimize operational friction. When the process operates smoothly, patients experience improved outcomes, and providers benefit from more efficient workflows.

These results reflect the actual benefits seen by institutions that have adopted self-scheduling platforms. For example, between January 1, 2019, and June 30, 2021, Johns Hopkins Community Physicians deployed automated self-scheduling through Epic's MyChart portal across more than 30 practices [11].

While digital tools have reshaped the healthcare experience, it's essential to pinpoint the specific moments when patients interact with these tools to better guide their journey. The use of self-scheduling increased from 4% to 15% of kept appointments. Meanwhile, patients who used self-scheduling had a 2.7% no-show rate, significantly lower than the 4.6% rate for agent-booked appointments. Although cancellations increased slightly (37.6% vs. 27.0%), the overall uptake improved efficiency and patient satisfaction.[11]

In addition to platform-based interactions, healthcare providers are finding value in more informal digital environments, namely, patient-led and provider-supported online communities.

In addition to structured platforms like clinic websites and patient portals, patients increasingly engage with less formal online communities. These spaces offer another layer of connection and influence that providers can't afford to ignore.

Benefits of Healthcare Providers Interacting with Online Communities

Provider-led online communities provide patients with a platform to connect, ask questions, and offer support to one another. These forums help people make better healthcare decisions by providing access to peer experiences, shared knowledge, and practical advice [12]. Some communities focus on chronic conditions and attract both patients and providers. This structure fosters open dialogue, builds confidence, and empowers participants to take more control over their care [13].

One well-known example is *PatientsLikeMe*, a disease-specific network that supports people with chronic illnesses. It connects users with others who have similar diagnoses and provides tools to track symptoms, explore treatments, and share progress [14]. Providers sometimes participate in these forums, using them to understand patient needs and contribute reliable information. The platform encourages collaboration and helps both patients and professionals learn from real-world experiences.

These examples demonstrate the value of integrating online communities into a comprehensive digital strategy. Providers who monitor or participate in these spaces can identify trends, address concerns, and build trust. A structured approach helps improve engagement and supports better outcomes while remaining compliant with privacy regulations such as HIPAA [15].

While community platforms foster engagement through shared experience, structured feedback systems offer measurable insights that help providers refine their digital strategy. While community forums reveal patterns in patient concerns, structured feedback tools give providers quantifiable data to improve care delivery and user experience.

Patient Feedback

Beyond social communities, individual feedback tools also provide important insights into patient preferences and needs. Digital feedback systems help healthcare staff by making it easier to collect, analyze, and apply patient input. Platforms like Press Ganey or built-in EHR survey modules can automate this process. Table 5.1 lists common digital feedback tools, example uses, and platform names. These tools include web-based surveys, patient portals, and SMS-based forms. Staff prefer these methods because they are quick to use, generate structured data, and integrate easily into clinical workflows [16].

Table 5.1 Types of digital feedback tools

Tool type	Example use	Common platform
QR survey	Post-visit impressions	Press Ganey, Typeform
Patient portal	Visit summaries, ratings	Epic MyChart, Cerner
SMS link	One-question NPS	CipherHealth, Feedtrail

Patients respond to digital tools in various ways. Some engage immediately, while others hesitate or prefer in-person feedback. These differences underscore the importance of designing tools that are accessible, secure, and easy to use.

When practices collect feedback consistently, they can spot problems early, adjust communication, and improve digital content. Real-world feedback can often reveal friction points that clinicians miss. For example, a BMC Health Services Research review found that patient feedback prompted over 100 changes across health facilities in Scotland, ranging from alterations to signage to improvements in service delivery. In one system, monitoring themes such as wait times, communication gaps, and patient comfort led to targeted updates in emergency and inpatient units, resulting in a reduction in service complaints within months.[1]

Analytics and feedback also help teams plan their digital strategy. By studying how patients respond, clinics can refine their messaging, enhance service delivery, and strengthen their relationships with the community. But collecting and using patient feedback responsibly requires careful attention to privacy and data protection. Collecting patient feedback is only valuable if organizations respond with operational changes. The next section focuses on integrating digital tools into workflows and staff routines.

Operational Integration and Staff Training

Digital tools only succeed when they are fully embedded into daily workflows. This includes the way front-desk staff check in patients, how nurses handle digital messages, and how physicians encourage portal use during visits. Without clear systems and consistent staff training, even the best-designed tools go unused or cause confusion.

Workflow mapping helps clarify these systems. It involves documenting how staff and patients move through each step of a visit, from appointment scheduling to checkout, and identifying where digital tools fit in. For example, if a patient schedules online but still receives a redundant call from the office, the workflow is broken. (For a sample template to guide this process, see Appendix 5.1).

Effective adoption starts with training. Each team member needs to understand their role in using digital tools and responding to patients. Front-desk staff should

[1] https://cipherhealth.com/blog/using-real-time-surveys-to-deliver-exceptional-patient-experiences-with-high-reliability/?utm_source=chatgpt.com. Accessed 7/28/25.

know how to troubleshoot patient portal access. Nurses should know where to log secure messages. Physicians should model digital behaviors by referencing portal content during appointments or encouraging feedback through QR codes. (For a comparison of leading digital feedback tools, including platform features and integration capabilities, see Appendix 5.3).

Physicians also play a leadership role. In many practices, they set the tone for whether digital tools are used consistently. When physicians use secure messaging, self-scheduling, and documentation tools themselves, staff are more likely to follow suit.

To support adoption

- Review each workflow step to identify where digital tools are used.
- Assign clear responsibilities for platform tasks like messaging, reminders, and data entry.
- Provide quick-reference guides and brief, role-specific training sessions. Appendix 5.2 provides a staff digital training checklist with role-specific items.
- Monitor key metrics such as tool usage rates, response times, and patient satisfaction.

Digital tools are not a side project. They are now part of patient care. Practices that treat them as core clinical infrastructure, like exam rooms or prescription pads, see better results and fewer staff frustrations.

Privacy and Security in Digital Healthcare

Maintaining patient trust depends on protecting private health data. HIPAA requires covered entities to meet three basic rules when using digital tools. They must secure any technology that stores protected health information (PHI), monitor and log access to PHI, and encrypt it to prevent unauthorized use [17].[2] These rules apply to websites, CRMs, email tools, and any connected applications used in patient outreach or recordkeeping. Figure 5.2 outlines core HIPAA security steps that apply across all digital engagement tools used in healthcare.

A robust cybersecurity strategy also encompasses regular software updates, employee training, and well-defined internal policies. Healthcare organizations should continuously monitor their systems to detect breaches early and respond quickly. They should also conduct regular audits to verify compliance and ensure that staff understand their responsibilities.

[2] https://www.forbes.com/sites/forbesbusinessdevelopmentcouncil/2021/03/12/the-use-of-technology-and-hipaa-compliance/?sh=20cefb9d1fff

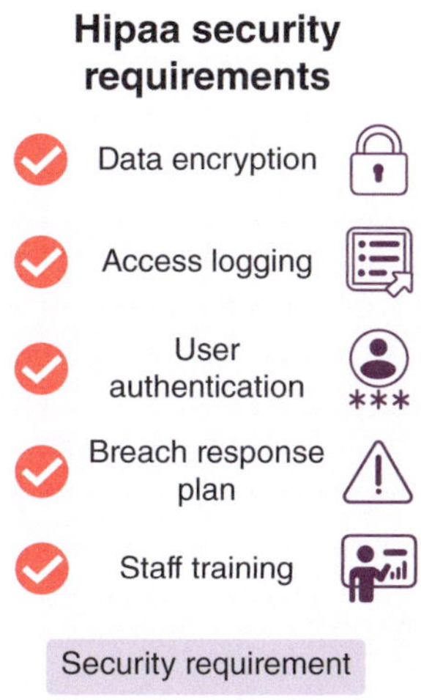

Fig. 5.2 HIPAA security requirements for digital data. (Quinn/ChatGPT)

Cyberattacks are becoming more frequent and sophisticated [18]. In 2024, 92% of healthcare organizations reported at least one cyberattack, up from 88% in 2023.[3,4] Healthcare facilities see the highest breach costs of any industry. In 2024, the average cost of a healthcare data breach came down to $9.77 million, compared to $10.93 million in 2023.[5]

Common threats include phishing emails, ransomware, and unpatched software vulnerabilities [18]. To reduce risk, healthcare providers must keep security protocols up to date and limit access to sensitive data based on job roles.

A well-structured privacy policy, visible on your website, reassures patients that their data is handled responsibly. This also includes clear messaging about consent and data use when collecting information through forms or patient portals.[6]

When providers take privacy seriously, patients are more likely to engage online and trust digital tools. With digital tools embedded in daily care, practices must also safeguard the data they generate. Protecting patient privacy is essential to sustaining trust and ensuring compliance in this increasingly connected environment.

[3] https://www.dialoghealth.com/post/healthcare-cybersecurity-statistics?utm_source=chatgpt.com. Accessed 7/29/25.

[4] https://www.proofpoint.com/us/resources/threat-reports/ponemon-healthcare-cybersecurity-report

[5] https://www.beckershospitalreview.com/healthcare-information-technology/cybersecurity/average-cost-of-healthcare-data-breach-by-year/?utm_source=chatgpt.com. Accessed 7/29/25.

[6] https://www.digitalguardian.com/blog/healthcare-cybersecurity-tips-securing-private-health-data

Conclusion

Digital tools have redefined how patients engage with healthcare—from first searches to follow-up care. When implemented thoughtfully, these tools improve access, streamline communication, and strengthen provider-patient relationships. But digital adoption also introduces new responsibilities around privacy, data security, and maintaining trust.

Practices that succeed in this space treat digital systems not as add-ons, but as integral parts of patient care. By embedding tools into workflows, training staff, listening to patient feedback, and prioritizing data protection, healthcare providers can deliver better outcomes while reducing friction across the patient journey.

Quick Checklist: Making the Digital Journey Work

☐ Are digital tools mapped into your daily workflows (scheduling, messaging, follow-up)?
☐ Does every staff role include training for digital tasks?
☐ Do you collect and apply patient feedback from portals, surveys, or SMS tools?
☐ Are online touchpoints (web, app, social media) monitored for engagement and usability?
☐ Is your patient data protected through encryption, audits, and HIPAA-compliant platforms?

Guiding patients through this evolving digital landscape builds more than convenience—it builds trust, loyalty, and lasting clinical performance.

Appendixes

Appendix 5.1: Sample Workflow Mapping Template

This sample workflow mapping template helps healthcare practices identify how digital tools are integrated into daily clinical routines. It can be used to analyze each step of the patient journey, assign responsibilities, and pinpoint inefficiencies. This structured approach supports smoother operations, improved patient experiences, and better use of digital platforms.

Workflow Step	Responsible Staff	Digital Tool / Platform	Time Estimate	Notes
Appointment Scheduling	Front Desk Staff	Epic MyChart	2 minutes	Confirmations auto-sent to patient email

Appendix 5.2: Staff Digital Training Checklist

Purpose

This checklist ensures that each staff member is trained to use digital tools appropriately in daily workflows. It supports consistency, improves patient experience, and strengthens HIPAA compliance.

Role	Training area	Completed (✓)	Notes
Front desk	Set up patient portal accounts and troubleshoot common log-in issues		Ensure printed guides are available
Front desk	Manage self-scheduling queues and flag duplicate appointments		Requires EHR access training
Nurse	Triage secure messages from portal and escalate clinical concerns		Train on message tagging and logging
Nurse	Use digital check-in systems to verify consent forms		Include HIPAA and scope-of-practice review
Physician	Demonstrate use of portal features during visits (e.g., lab review, refills)		Reinforces digital adoption by patients
All staff	Communicate digitally using HIPAA-compliant platforms only		Include refresher on breach reporting
All staff	Review patient feedback and escalate service-related comments		Training includes tone and escalation paths

Appendix 5.3: Digital Feedback Tool Comparison Table

Purpose

This table compares leading digital feedback platforms, with attention to healthcare-specific needs like EHR integration, privacy, and usability. Selection should be based on workflow compatibility and patient population.

Tool	Feedback method	HIPAA-compliant	EHR integration	Analytics features	Language support
Feedtrail	SMS/email	Yes	Yes	Real-time dashboards	English, Spanish
Typeform	Web forms	With BAA	Limited	Response trends and NPS	Multilingual
CipherHealth	SMS/email	Yes	Yes	Automated trend reports	English, Spanish, others
Press Ganey	Web, QR codes	Yes	Full integration	Patient satisfaction indexing	English
Qualtrics	Web/kiosk	Yes	API available	Custom dashboards + alerts	Multilingual

Notes

- "With BAA" indicates HIPAA compliance is possible only after signing a Business Associate Agreement
- Language support is essential for accessibility and equity
- Use internal IT and compliance teams to vet integration and encryption standards

References

1. Blease C, Kharko A, Bernstein MH, Gaeta S, DesRoches C. Digital health and patient experience: an integrative review. J Patient Exp. 2021;8:23743735211054324.
2. Kruse CS, Krowski N, Rodriguez B, Tran L, Vela J, Brooks M. Telehealth and patient satisfaction: a systematic review and narrative analysis. BMJ Open. 2022;7(8):e016242.
3. Shah SGS, Liebovitz DM. It takes two to tango: engaging patients and providers with portals. JMIR Med Inform. 2017;5(2):e43.
4. American Medical Association (AMA). Digital health research: physicians' motivations and requirements for adopting digital health. 2022. https://www.ama-assn.org/system/files/ama-digital-health-study.pdf
5. Agarwal R, Gao GG, DesRoches C, Jha AK. Research commentary – the digital transformation of healthcare: current status and the road ahead. Inf Syst Res. 2020;21(4):796–809.
6. Mesko B, Győrffy Z, Kollár J. Digital literacy in the medical profession. J Med Internet Res. 2017;19(3):e91.
7. Davies D, Thomas K, Rashid R. Mapping digital patient pathways to improve care access and efficiency. Digital Health. 2023;9:1–10.
8. Alami H, Gagnon MP, Wootton R, Fortin JP, Zanaboni P, Lesage A. Exploring the barriers to the adoption and implementation of telemedicine services in chronic disease management. J Med Internet Res. 2020;22(6):e13387.

9. North F, Crane SJ, Stroebel RJ, Chaudhry R, Ebbert JO, Ytterberg KL. Patient portal use and health information seeking behavior. JMIR Med Inform. 2019;7(2):e11348.
10. Lyles CR, Sarkar U, Osborn CY. Getting a technology-based diabetes intervention ready for implementation in clinics: perspectives of staff and patients. J Am Med Inform Assoc. 2020;27(2):294–302.
11. Woodcock E, Sen A, Weiner J. Automated patient self-scheduling: case study. J Am Med Inform Assoc. 2022;29(9):1637–41. https://doi.org/10.1093/jamia/ocac087. PMID: 35652165; PMCID: PMC9382371
12. Smailhodzic E, Hooijsma W, Boonstra A, Langley DJ. Social media use in healthcare: a systematic review of effects on patients and on their relationship with healthcare professionals. BMC Health Serv Res. 2016;16:442.
13. van Uden-Kraan CF, Drossaert CH, Taal E, Seydel ER, van de Laar MA. Self-reported differences in empowerment between lurkers and posters in online patient support groups. J Med Internet Res. 2008;10(2):e18.
14. Frost J, Massagli M. Social uses of personal health information within PatientsLikeMe, an online patient community: what can happen when patients have access to one another's data. J Med Internet Res. 2008;10(3):e15.
15. Ventola CL. Social media and health care professionals: benefits, risks, and best practices. Pharm Ther. 2014;39(7):491–520.
16. Fung BN, Hodgson D, Small N, Rogers A, Sanders C. Implementing a digital patient feedback system: an analysis using Normalisation Process Theory. BMC Health Serv Res. 2020;20:337. https://doi.org/10.1186/s12913-020-05234-1.
17. Office for Civil Rights (OCR). Summary of the HIPAA security rule. U.S. Department of Health & Human Services. 2022. https://www.hhs.gov/hipaa/for-professionals/security/index.html
18. Health Sector Cybersecurity Coordination Center (HC3). Health industry cybersecurity practices: managing threats and protecting patients. U.S. Department of Health & Human Services. 2023. https://www.hhs.gov/sites/default/files/health-industry-cybersecurity-practices.pdf

Chapter 6
Advanced Strategies for Managing Online Reputation

A good reputation is more valuable than money.—Publilius Syrus

What Is Online Reputation?

An online reputation reflects the collective impression patients and the public form based on digital content. This includes reviews on sites like Yelp and Healthgrades, comments on social media, and mentions in online forums. A strong digital presence can influence whether a patient chooses one provider over another [1].[1,2] Managing that presence requires more than passively accumulating reviews. Providers should monitor feedback regularly, respond to both praise and criticism, and correct outdated or inaccurate listings. Encouraging satisfied patients to leave reviews and addressing complaints quickly and respectfully help shape a reputation built on trust, transparency, and responsiveness [2].

For example, a dental practice observed an increase in patient appointments after implementing a strategy to encourage online reviews from satisfied patients. They also made a point to respond thoughtfully to any negative reviews, which not only improved their online reputation but also demonstrated their commitment to patient satisfaction.[3]

Through diligent management and strategic engagement, healthcare providers can significantly enhance their online reputation, leading to increased trust, credibility, and a higher influx of patients.

[1] https://www.demandhub.co/articles/reputation-management-for-doctors/. Accessed 3/10/24.

[2] https://www.forbes.com/sites/forbesagencycouncil/2022/11/15/a-guide-to-healthcare-online-reputation-management/?sh=71c095a84d78. Accessed 3/10/24.

[3] https://www.doctible.com/guide/online-reputation-management-orm-guide-for-healthcare-professionals. Accessed 3/10/24.

C. Quinn, *The Digital Practice*, https://doi.org/10.1007/978-3-032-07745-5_6

Why It Matters to Physicians

Managing your online reputation is crucial for healthcare providers, as it has a significant impact on patient trust and satisfaction. A robust online presence, marked by positive reviews and active engagement, reassures potential patients of the quality of care they can expect, which is particularly important in the healthcare sector where personal stakes are high.[4] Moreover, a positive online reputation serves as a magnet for new patients, making it a critical factor in a healthcare provider's ability to attract business in the digital era.[5] As mentioned in previous chapters, the majority of healthcare consumers now research healthcare information and healthcare providers online before making an appointment [3]. Prospective patients often rely on online reviews from strangers when choosing a healthcare provider. A dentist rated at 4.7 stars with numerous positive testimonials is far more appealing than a competitor down the road with a 2 star rating.

Over the past two decades, the Internet has evolved from a modest collection of websites to a vast network comprising millions of local businesses, encompassing a wide range of medical practices, including those of doctors, specialists, dentists, podiatrists, chiropractors, therapists, treatment centers, and wellness professionals such as physical therapists. Patients use the Internet to find these providers. A 2022 survey revealed that 77% of patients use online reviews as their first step in selecting a healthcare provider.[6]

In the past, patients typically searched for providers using specific queries such as "physical therapists in Trenton, NJ." These searches often led to third-party listings on platforms like Healthgrades, Vitals, or RateMDs.

Today, most users conduct location-based searches using generic phrases such as "physical therapists near me." Search engines now automatically detect geographic location and prioritize results accordingly. This shift has dramatically changed how patients discover providers.

This type of search has become so popular that over the last decade, the frequency of "near me" searches has skyrocketed by over 900%.[7],[8]

[4] https://www.forbes.com/sites/forbescommunicationscouncil/2022/01/12/why-does-online-reputation-matter-for-healthcare-practices/?sh=405e152d2564. Accessed 3/10/24.

[5] https://rater8.com/blog/healthcare-reputation-management/. Accessed 3/10/24.

[6] https://b2b.healthgrades.com/insights/blog/online-reviews-impact-how-patients-select-hospitals-doctors/#:~:text=CONSUMER%20TRENDS%20SHOW%3A&text=80%25%20of%20consumers%20trust%20online,in%20finding%20a%20new%20doctor. Accessed 7/29/25.

[7] https://www.thinkwithgoogle.com/marketing-strategies/app-and-mobile/near-me-searches/. Accessed 3/10/24.

[8] https://seoprofy.com/blog/local-seo-statistics/?utm_source=chatgpt.com. Accessed 7/28/25.

Vignette

Dr. Hirsh had never paid much attention to his practice's online reviews. He assumed word-of-mouth referrals and his spotless record were enough. Then one day, a new patient canceled after reading a two-star review that described her front-desk staff as "rude and dismissive." That same week, a colleague mentioned that his patient load had doubled after he started asking satisfied patients to leave Google reviews. Dr. Hirsh searched for his own name and found outdated contact information, inconsistent ratings, and no clear way for patients to schedule appointments. His digital reputation didn't reflect the care he delivered every day.

Like many physicians, Dr. Hirsh had invested his efforts in improving clinical quality but overlooked the digital signals that shape public perception. Managing your online reputation now requires more than excellence in care; it demands a deliberate, ongoing strategy.

If you've never Googled your own name, now is the time.

Google Business Profile Optimization

Google's algorithm often displays the top local matches in a prominent section on the results page known as the "local 3-pack." This area includes a map and three high-ranking business profiles, making it the most visible placement for practices. For healthcare providers, appearing in the 3-pack increases the likelihood of patient inquiries, appointment bookings, and improved digital visibility. For physicians, securing a spot in the 3-pack is a powerful digital marketing advantage; it can significantly increase patient inquiries, appointment bookings, and local visibility.

Rankings in this section depend on three factors: relevance to the query, proximity to the searcher, and online prominence. Prominence includes the quality and quantity of online reviews, accuracy of contact details across listings, and completeness of the Google Business Profile.

Appearing in the 3-pack also boosts credibility, especially for patients searching for terms like "primary care near me" or "dermatologist in [city]," where trust and accessibility are key decision factors. The quality and accuracy of a healthcare practice's website affect the practitioner's reputation.

While optimizing your Google Business Profile is essential, it's equally important to monitor third-party listings on healthcare-specific platforms like Healthgrades, Zocdoc, and Vitals.

Callout Box:Beyond Google: Monitor Healthcare Directories Patients Use

Patients often rely on medical-specific platforms like Healthgrades, Zocdoc, Vitals, and WebMD to compare providers, verify insurance, and book appointments. Many of these sites create listings automatically using licensing boards, health systems, or insurance data, so your profile may exist whether you've claimed it or not.

To protect your reputation and ensure patients find accurate information

- **Search** your name and practice across major healthcare directories.
- **Claim** and verify each listing. Check address, phone, specialty, insurance participation, and scheduling links.
- **Correct** duplicates or merged profiles, especially in shared office spaces.
- **Monitor monthly** for unauthorized changes or new patient reviews.

If these listings show outdated or conflicting information, patients may book with the wrong provider or choose not to book at all.

Before you start asking patients for reviews or responding to online feedback, make sure the foundation of your digital presence is solid. A well-optimized Google Business Profile attracts more attention, but if the information on that profile is wrong or outdated, it can cause confusion and damage trust. That's why accuracy isn't optional. It's essential.

Accuracy Is Important

Maintaining accurate and up-to-date online information about the practice and its healthcare providers is crucial for several reasons. First, it ensures that patients have access to reliable and current data, which is essential for making informed health decisions. As highlighted by Tonsaker et al. [4], online health information can significantly enhance patients' knowledge, competence, and engagement in decision-making.[9] But when listings are outdated or incorrect, such as listing the wrong location, phone number, or services, they do more than confuse patients. In high-stakes specialties like surgery or obstetrics, misinformation can lead to delayed care, missed procedures, or inappropriate referrals. These missteps carry real liability risks for providers, especially if a patient suffers harm due to relying on faulty directions or unverified service availability. From a legal standpoint, failing to correct incorrect digital listings may be construed as negligence if it results in patient harm.

[9] https://www.ncbi.nlm.nih.gov/pmc/articles/PMC4020634/

A medical clinic, for example, learned that patients were arriving at the wrong location for scheduled procedures because its Google Business Profile had been incorrectly merged with that of a nearby facility. The clinic had never claimed or verified its profile, which had been auto-generated from third-party data. As a result, Google's algorithm combined it with another institution's listing. This led to patients showing up for scheduled surgeries at the wrong address and sometimes at an entirely unrelated hospital. The confusion caused delays in care and created unnecessary stress for both patients and staff.[10]

This case demonstrates that when local listings are inaccurate or unclaimed, patient trust and safety can be compromised, particularly in healthcare, where arriving at the correct location is crucial for timely treatment.

Keeping your online profiles accurate isn't a one-time task. Use this checklist to perform regular audits across all platforms and prevent errors that can disrupt care, confuse patients, or damage your reputation.

Checklist: Accuracy Audit Tips
Use this checklist to ensure your online presence reflects accurate, up-to-date information across all platforms:

☐ **Claim and Verify All Listings**
Secure your profiles on Google, Healthgrades, Zocdoc, Vitals, WebMD, and any insurer directories.

☐ **Check Contact and Location Details**
Verify that your clinic name, phone number, address, and office hours are correct and consistent across all listings and maps.

☐ **Confirm Insurance Participation**
Ensure health plan information is current on insurer websites and patient-facing platforms.

☐ **Review Provider Credentials and Services**
Confirm that your bio, specialties, degrees, and offered services match what's available in your practice.

☐ **Audit Scheduling Links and Forms**
Click through all online scheduling buttons and forms—on your site, review platforms, and patient portals—to confirm they route correctly.

☐ **Validate Map Pins on Mobile and Desktop**
Check Google Maps and Apple Maps to ensure the marker lands on the correct building or entrance.

☐ **Eliminate Duplicates and Merged Listings**
Correct or merge profiles that confuse patients, especially in shared buildings or group practices.

[10] https://chrisabraham.com/blog/when-google-sends-your-patients-to-the-wrong-place-the-high-stakes-chaos-of-an-unclaimed-or-mislabeled-google-business-profile?utm_source=chatgpt.com. Accessed 7/28/25.

- ☐ **Update Social Media and EHR-Linked Pages**
  Ensure business pages on Facebook and LinkedIn, as well as any EHR-published portals (e.g., athenahealth, eClinicalWorks), are current.
- ☐ **Assign a Point Person**
  Designate a staff member or admin to conduct the audit regularly and document updates.
- ☐ **Monitor Quarterly (or Monthly for High-Volume Clinics)**
  Schedule recurring reviews to catch unauthorized changes or system errors before patients are affected.

A well-managed online reputation establishes healthcare providers as credible and authoritative sources of medical information and care. Engaging with patients online, responding to reviews, and sharing valuable health information contribute to a professional image that resonates with both existing and potential patients.[11]

This strategy not only helps retain patients but also enhances the provider's credibility and authority as a source of medical information and care. For instance, a family practice clinic saw a significant improvement in its online reputation and an increase in patient appointments by actively soliciting feedback from patients using reputation management tools like BrightLocal and Birdeye and by promptly and personally addressing negative reviews. This example underscores the tangible benefits of effective online reputation management in the healthcare industry.

Generating and Managing Online Reviews

In the realm of digital marketing for healthcare providers, proactive online review management stands as a cornerstone strategy. This approach entails actively monitoring and responding to both positive and negative reviews across various platforms. Engaging with reviews in a timely and empathetic manner not only demonstrates a healthcare provider's commitment to patient satisfaction but also significantly influences the perception of potential patients. (For a comparison of HIPAA-compliant review platforms and their features, see Appendix 6.1.) Table 6.2 in Appendix 6.1 compares leading patient review tools and their HIPAA features, so you can pick the right platform.

The first step involves assessing the current online reputation by closely monitoring reviews on popular platforms. Tools like BrightLocal, Birdeye, and Grade.us are invaluable in facilitating the process of sending review requests to recent patients, thereby encouraging a steady flow of feedback.

[11] https://www.demandhub.co/articles/healthcare-reputation-management/#:~:text=Benefits%20of%20Reputation%20Management&text=Boost%20patient%20satisfaction%20and%20trust,proactively%20and%20improving%20patient%20experience. Accessed 3/10/24.

A review request refers to any intentional outreach that encourages patients to leave an online review after their visit. This may include automated text messages (SMS), emails with embedded links to review sites, in-app notifications from patient portals, or even verbal reminders at checkout. Many practices use reputation management platforms to automate these requests shortly after the appointment ends.

A positive online presence can alleviate patient anxieties, instill confidence, and, ultimately, sway their decision-making process in favor of the healthcare provider.

Managing Negative Feedback

Negative feedback is inevitable, even for the most skilled healthcare providers. What matters is how your practice responds. A prompt, respectful reply can demonstrate that you take concerns seriously and are committed to resolving problems. Turning negative feedback into positive outcomes is a crucial skill for healthcare providers in the digital age.[12]

The most effective responses acknowledge the patient's experience without revealing protected health information. A simple apology, paired with a request to speak offline, signals professionalism and a desire to improve.

Handled well, even a negative review becomes an opportunity. Patients want to see that a practice listens and adapts. A thoughtful response builds trust, not only with the reviewer but with every potential patient reading your reply. Over time, this approach strengthens your credibility and shows that your commitment to care extends beyond the exam room.

Use the guide below (Table 6.1) to ensure your public responses to patient reviews stay professional, respectful, and fully HIPAA-compliant.

Best Practices for HIPAA-Compliant Review Responses

For instance, in response to a negative review concerning long wait times at his office, the doctor took immediate and thoughtful action to address the patient's concerns. Recognizing the importance of patient satisfaction and the impact of wait

Table 6.1 Do's and don'ts of HIPAA-compliant review responses

What to do	What to avoid
Use general, respectful language	Mentioning patient names, dates, or procedures, even vaguely
Invite the reviewer to contact your office directly	Attempting to resolve complaints in the public reply
Thank them for their feedback	Acknowledging or denying their patient relationship

[12] https://www.medicaleconomics.com/view/online-reviews-are-becoming-more-important-to-patients-in-choosing-their-care-how-to-manage-your-online-reputation-in-health-care Accessed 3/10/24

times on the overall patient experience, the doctor responded promptly to the review, demonstrating attentiveness and a commitment to resolving the issue. The response included an apology for the inconvenience caused and an acknowledgment of the value of the patient's time.

The doctor then explained the steps being taken to improve wait times, such as implementing a new appointment scheduling system designed to reduce bottlenecks and ensure a smoother patient flow through the office. Additionally, the doctor mentioned the introduction of a more efficient check-in process and the hiring of additional staff during peak hours to better manage patient flow.

To further demonstrate commitment to patient care and satisfaction, the doctor invited the patient to discuss their concerns directly, offering a personal contact method. This open line of communication not only showed the doctor's dedication to improvement but also reinforced the practice's commitment to transparency and patient-centered care. By taking these actions and communicating them effectively, the doctor turned a negative review into an opportunity to showcase the practice's proactive approach to addressing patient feedback and making tangible improvements.

This strategy not only helps attract more patients but also outshines the competition by building trust and credibility with potential patients, thereby underscoring the importance of a meticulously managed online reputation in the healthcare sector.

Responding to patient reviews, especially negative ones, can be an essential part of your digital reputation strategy; however, it also presents serious legal risks if not handled properly.

Responses and Risks

Under the Health Insurance Portability and Accountability Act (HIPAA), covered entities (like physician practices and clinics) are prohibited from disclosing any protected health information (PHI) in public forums, including online review platforms such as Google, Yelp, or Healthgrades.

This means

- You cannot confirm or deny that someone is a patient.
- You cannot reference their diagnosis, treatment, or visit details.
- Even if the patient discloses their own health information in the review, you may not reciprocate with any acknowledgment or detail that could confirm their identity or relationship to your practice. Even vague references like "your surgery last week" can violate HIPAA because they confirm details about the patient's care, which qualifies as protected health information (PHI) even if no name is mentioned.

Violations of this policy, even if unintended, can result in civil penalties of up to $50,000 per violation and significant reputational harm.

Example of a Compliant Response

Thank you for your feedback. We take concerns seriously and would appreciate the opportunity to discuss this further. Please contact our office directly so we can address your experience.

Example of a Noncompliant Response (Do Not Use)

We're sorry your root canal took longer than expected, but you missed your 1:30 appointment, which caused the delay.

Even well-intentioned replies like this can expose PHI and constitute a HIPAA violation.

Conclusion

In today's healthcare environment, online reviews are no longer optional. They are a vital part of your practice's reputation, patient trust, and local visibility. Leveraging review tools, optimizing your Google Business Profile, and responding thoughtfully to patient feedback can enhance your online presence and foster long-term loyalty. However, it is equally important to navigate this space with care: physicians and practice staff must remain vigilant about HIPAA compliance when engaging online. By combining a proactive digital strategy with ethical and legal safeguards, you can strengthen your brand, protect patient privacy, and ensure your practice thrives both online and in person. Review management should remain an internal responsibility or be assigned only to trusted partners who understand HIPAA obligations, never outsourced blindly to third-party marketers unfamiliar with healthcare regulations.

Appendix 6.1 (Table 6.2)

Table 6.2 Top patient review tools for medical practices

Tool	HIPAA-compliant	Unique features	Best for
Birdeye	Yes	• Integrates with eClinicalWorks, athenahealth, Dentrix• AI sentiment analysis• Text-based review requests	Multispecialty practices or large medical groups
Podium	Yes (with BAA)	• SMS-based review collection• Webchat + Google integration• EHR/PMS integration via API/Zapier	Mobile-first practices aiming for high review volume
Swell	Yes	• Integrates with ModMed, Open Dental• Auto-sends review requests• Google/Facebook review targeting	Specialists like dermatology, dental, orthopedics
Reputation.com	Yes	• Enterprise-grade• Integrates with Epic, Cerner, athenahealth• Advanced analytics and AI review insights	Health systems, hospitals, or large provider networks
Doctible	Yes	• Online scheduling + appointment reminders• Payment reminders• SMS/email review requests	Independent practices needing all-in-one patient communication tools
PatientPop (Tebra)	Yes	• SEO optimization• Integrates with EHRs and websites• Visibility tracking dashboard	Growth-oriented practices investing in digital presence
Weave	Yes	• VoIP + texting + scheduling• Reviews triggered by call events• Mobile app for staff	Practices wanting unified communication and review features

HIPAA Compliance Reminder

Ensure your review tool provides a Business Associate Agreement (BAA) and avoids exposing protected health information (PHI) in public-facing reviews.

Please Note Tool descriptions are for informational purposes only and do not constitute endorsements.

References

1. Healthgrades, Healthgrades 2022 online patient trends report. 2022. https://marketing.healthgrades.com/rs/187-VHH-001/images/HG-2022-Online-Patient-Trends-Report.pdf
2. Physicians Practice. 5 ways to boost your online reputation management strategy. 2022. https://www.physicianspractice.com/view/5-ways-to-boost-your-online-reputation-management-strategy. Accessed 28 Jul 2025.
3. Hesse BW, Moser RP, Rutten LJ. Surveys of physicians and electronic health information. N Engl J Med. 2010;362(9):859–60.
4. Tonsaker T, Bartlett G, Trpkov C. Health information on the Internet: gold mine or minefield? Can Fam Physician. 2014;60(5):407--8. PMID: 24828994; PMCID: PMC4020634.

Chapter 7
Mastering Local SEO for Medical Practices

What Is SEO, and Why Is It Important?

SEO, or search engine optimization, is the process of optimizing your website to attract more visitors through organic search results. In the competitive healthcare industry, SEO can be the differentiator that sets a medical practice apart from its competitors. By achieving higher rankings, healthcare providers can capture a larger share of online searches, attracting more patients to their practice than their competitors.[1]

A study of search engine activity done in October of 2023 showed that the first five listings in search results get 68% of the clicks.[2] Healthcare providers interested in capturing these coveted spots adjust their websites so that search engines preferentially choose them. This is the reason for SEO.

Not only do top rankings improve your business's visibility, but they can be seen as endorsements from search engines like Google. When a healthcare website appears at the top of search results, it signals to users that the site is a credible and authoritative source of information. This trust is crucial in the healthcare sector, where the accuracy of information can directly impact patient health.[3]

You can improve your SEO by incorporating specific keywords, improving site speed, ensuring mobile-friendliness, and creating quality content that meets the needs of your audience.[4] To put these strategies into action, your first priority should be claiming and correcting your business listings on platforms patients already use.

---

[1] https://healthcaresuccess.com/blog/healthcare-marketing/healthcare-seo-guide.html. 3/5/24.

[2] https://firstpagesage.com/reports/google-click-through-rates-ctrs-by-ranking-position/. Accessed 3/5/24.

[3] https://www.forbes.com/sites/forbesagencycouncil/2022/01/20/why-seo-for-medical-practices-is-important-for-getting-more-patients-consistently/?sh=32705ec43e75. Accessed 3/5/24.

[4] https://www.siteimprove.com/glossary/healthcare-seo/. Accessed 3/4/24.

C. Quinn, *The Digital Practice*, https://doi.org/10.1007/978-3-032-07745-5_7

Once you understand the importance of SEO, the first step in applying it to your practice is claiming and correcting your online business listings.

Leveraging Local Business Directories

Ensure your medical practice is accurately listed in local business directories such as Google Business Profile, Yelp, and Healthgrades. Many third-party directories automatically create business profiles using publicly available data, but these listings often contain errors and must be manually claimed and corrected by the practice to ensure accuracy. Each profile should include the same name, address, phone number (NAP), and list of services. Consistency across these listings helps search engines verify your legitimacy and improves your ranking in local search results. When your information is reliable and complete, search engines are more likely to display your clinic to patients searching for care in your area. This improves your visibility, builds trust, and increases appointment inquiries from new patients.[5,6]

See Fig. 7.1 for a step-by-step visual guide to identifying and correcting broken business listings across major platforms.

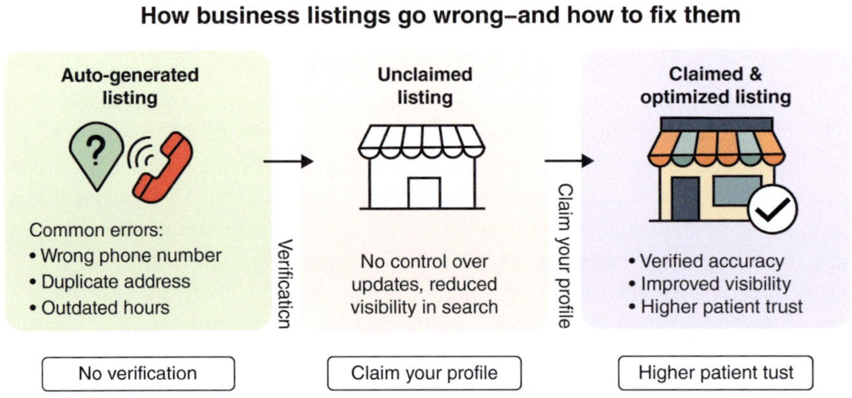

Fig. 7.1 Fixing broken business listings: a visual guide (Quinn/ChatGPT)

[5] https://blog.hubspot.com/marketing/local-seo. Accessed 3/11/24.

[6] https://brainchildstudios.com/blog/local-seo/10-ways-to-optimize-your-website-for-local-seo/. Accessed 3/11/24.

Moreover, being listed in these directories also provides direct avenues for potential patients to discover your practice beyond search engine results, further increasing your online visibility and patient reach. This comprehensive online presence is crucial for improving your practice's SEO and attracting more patients in the digital age. Once your listings are accurate and consistent, the next step is refining the language patients use to find your services—this is where keyword strategy comes in. After you've ensured your business listings are consistent and complete, the next priority is helping patients and search engines find your website using the right words. That's where a strong keyword strategy comes into play.

Keywords

Incorporating local keywords into your website's content, titles, meta descriptions, and URLs significantly enhances your visibility in local search results. A meta description is a short summary of a webpage that appears under the title in search engine results. It helps users understand what the page is about before they click, and it can influence whether someone visits your site.

For instance, rather than using broad terms such as "family doctor," specifying "family doctor in [City Name]" or "pediatrician near [Neighborhood Name]" can make your practice more discoverable to potential patients in your vicinity.[7,8] This targeted approach to search engine optimization (SEO) draws in traffic that is not only relevant but also more likely to engage with your services.

For example, a dermatologist can optimize their site to appear for searches like "acne treatment" or "skin care specialist," thereby attracting individuals actively seeking these specific services.[9] Such strategic use of SEO not only enhances the user experience by connecting patients directly with the medical services they require but also boosts the likelihood of converting website visitors into patients. This makes SEO a crucial element of digital marketing for healthcare providers, as it bridges the gap between patients and the healthcare services they are searching for.[10,11]

Choosing the right keywords is only the beginning. You also need to structure your website content so search engines can recognize and reward it. That's where metadata comes in.

---

[7] https://www.webfx.com/industries/health/hospitals-healthcare/local-seo/. Accessed 3/11/24.

[8] https://digitalismedical.com/blog/local-seo-for-healthcare/. Accessed 3/11/24.

[9] https://www.semrush.com/blog/healthcare-seo/. Accessed 3/5/24.

[10] https://www.getfoundquick.com/5-reasons-why-seo-is-important-for-the-healthcare-industry/. Accessed 3/4/24.

[11] https://simplifiedseoconsulting.com/why-is-medical-seo-important-healthcare-seo/. Accessed 3/4/24.

Metadata and Title Tags

Metadata, or “data about data,” provides search engines with background information about your site’s content. This includes details like the page’s description, keywords, author of the document, and more, which are not visible on the webpage itself but can be crucial for search engine optimization (SEO).[12]

Title tags, also known as meta titles, are a specific type of metadata that tells search engines and users the title of a webpage. This title is typically displayed on search engine results pages (SERPs) as the clickable headline for a given result and is essential for both SEO and social sharing. The title tag of a webpage is meant to be an accurate and concise description of a page’s content, making it a critical element for both user experience and search engine ranking.[13] For example, a strong title tag might be “Pediatrician in Albany | Main Street Pediatrics.” This format clearly identifies the specialty, location, and practice name, helping both users and search engines understand the page content.

Metadata and title tags are crucial for creating a strong initial impression on potential visitors. For local SEO optimization, it’s essential to incorporate the city’s name into your page title tags, alongside the service you provide. This approach serves a dual purpose: it signals to search engines your business location and explicitly communicates to users the services you offer in that specific area.

Even perfectly worded content won’t help if your site takes too long to load. Speed is now a ranking factor and a patient satisfaction issue.

Speed and Impatience

A slow website drives patients away.[14] Studies show that most users abandon a site if it takes more than a few seconds to load [1]. Search engines factor this into rankings because they prioritize user experience [2]. Google explicitly uses page loading speed as a key ranking factor, especially as measured by Core Web Vitals in 2025. Sites that load slowly are penalized in search results, while fast-loading sites are favored.[15] This applies to all sectors, including healthcare. Faster sites are easier to navigate, load more reliably on mobile, and encourage longer engagement.[16,17]

---

[12] https://www.wordstream.com/meta-tags. Accessed 3/11/24.

[13] https://www.semrush.com/blog/meta-tag/. Accessed 3/11/24.

[14] https://neilpatel.com/blog/does-speed-impact-rankings/. Accessed 3/5/24.

[15] Backlinko: https://backlinko.com/google-ranking-factors. Accessed 8/8/25.

[16] https://www.cloudways.com/blog/impact-of-website-speed/#:~:text=SEO%20and%20Website%20Speed,-Website%20speed%20is&text=A%20faster%20website%20is%20more,its%20emphasis%20on%20website%20speed. Accessed 8/8/25.

[17] Growtha https://growtha.com/blog/improving-page-speed-for-medical-websites. Accessed 8/8/25.

Your website's loading speed depends on several factors: server performance, file sizes, image compression, and hosting setup. It is not simply a matter of your own Internet connection. A well-optimized site loads quickly across devices and makes it easier for search engines to scan your content. Search engines favor faster sites because they offer better patient experience and reduce abandonment rates.[18]

To stay competitive, aim for a page load time of 2.5 s or less on desktop and under 8.6 s on mobile. These benchmarks are based on an analysis of the top 100 websites globally and reflect patient expectations for responsiveness [3].

For practices offering telehealth services, connection speed matters. The Federal Communications Commission recommends a minimum of 25 Mbps download and 3 Mbps upload for solo physician offices that handle basic digital functions like scheduling, messaging, and web browsing.[19] For telemedicine, download speeds of at least 10 Mbps and upload speeds of 5 Mbps help maintain clear, stable video sessions between patients and providers [4].[20] Improving site speed isn't only a technical upgrade, but it directly improves your visibility in search results and reduces patient drop-offs.

Site speed affects how well your site ranks today, but search behavior is already evolving. With the rise of AI-generated summaries in search results, your content must now be optimized not only for visibility but also for machine readability. This shift demands a new layer of SEO readiness.

Preparing for AI-Powered Search Results

Google's new *Search Generative Experience* (SGE) is changing how patients find doctors online. Instead of displaying a traditional list of domain links, Google now often shows an AI-generated summary first. These summaries appear before standard results and offer direct answers to user queries. For example, when patients look for providers, SGE may provide a snapshot that outlines specialties, location relevance, and contact info, without requiring users to click through to individual websites. This shift changes how people discover doctors online and emphasizes the importance of presenting content that AI can easily interpret.

As shown in Fig. 7.2, adding the FAQ schema changes a standard search listing into a richer, more interactive result that draws more clicks and improves patient engagement.

This means your practice website must do more than rank well; it needs to be written and structured in a way that makes it easy for Google's AI to understand and

[18] https://searchendurance.com/page-speed-seo/?utm_source=chatgpt.com. Accessed 7/30/25.

[19] https://www.healthit.gov/faq/what-recommended-bandwidth-different-types-health-care-providers. 3/5/24.

[20] https://blog.curogram.com/telehealth-tech-requirements. 3/5/24.

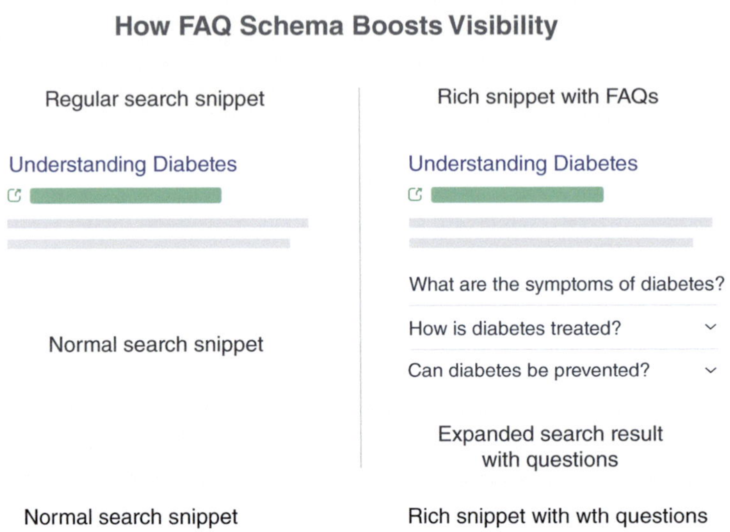

Fig. 7.2 This infographic shows how adding FAQ schema to your website transforms a basic search listing into a richer, more visible result with interactive questions that boost user engagement. (Quinn/ChatGPT)

quote your content accurately. If you don't make these adjustments, your practice may become invisible, even if you have excellent SEO in place.

Here's how to prepare:

1. **Use Structured Data Tags to Highlight Key Facts**

   Structured data tags (also called schema markup) are bits of code that tell Google what your content means. For example, you can tag your name as a "physician," your location as a "clinic address," or your insurance list as "acceptedHealthPlan." These tags help Google connect your website with relevant search terms like "primary care doctor near me who takes Aetna."

   How to Do This Without Writing Code.

   You don't need to learn HTML. Use free tools like the following:

- Google's Structured Data Markup Helper
- Schema App

- These tools walk you through the process step-by-step. You paste in your website URL, highlight essential items (like your name, address, or services), and the tool generates the structured data code. Your web developer can then paste this code into your website. If you use a platform like WordPress, plugins such as **Yoast SEO** or **Rank Math** can do this for you with a few clicks.

2. **Create a Clear and Patient-Friendly FAQ Section**

   Write answers to common questions patients type into Google. This helps SGE recognize your site as a helpful source. Keep the answers short, clear, and conversational. Examples

- "What is a lipid panel and why is it ordered?"
- "How do I schedule an appointment?"
- "Do you accept Blue Cross Blue Shield?"
- "Is parking available at your clinic?"

Put these FAQs directly on your service pages. For example, under your cardiology service page, include the following:

Q: How do I prepare for a stress test?
A: Wear comfortable shoes, avoid caffeine for 24 hours, and follow your doctor's instructions about medications.

3. **Answer Specific Medical Questions Using Natural Language**
   Write content the way you speak to patients. Instead of using jargon or writing in abstract, answer questions like the following:

- "What are the symptoms of high blood pressure?"
- "How often should I get a colonoscopy?"
- "What's the difference between urgent care and the ER?"

Google favors direct, helpful answers written in a tone patients understand.

4. **Earn Backlinks from Trusted Health Sources**
   Ask to be listed on local hospital referral pages, health directories, or professional society websites. Suppose a local news site features you in a health article, even better. These links back to your site tell Google that your content is reliable.

These minor technical updates, structured tags, natural phrasing, and backlinks add up to a bigger reward: being featured directly in AI-generated search summaries.

Bottom Line

Even if you've already invested in SEO, your website needs to speak the language of AI by using structured data and patient-centered phrasing. Structured data, patient-friendly FAQs, and medically accurate content increase the chances that Google's AI will feature your practice in its new summary results. You don't need to be a programmer; you just need the right tools and a bit of help from your web support team. See Appendix 7.1 for the Physician's AI SEO Readiness Checklist.

Conclusion

Mastering local SEO is crucial for healthcare providers aiming to enhance their online visibility and attract more patients. By incorporating local keywords, ensuring accurate listings in business directories, and effectively managing online reviews, medical practices can significantly improve their search engine rankings. Assigning a trusted staff member or web partner to monitor local SEO ensures that efforts are sustained over time.

This strategy not only makes it easier for potential patients to find your services but also establishes your practice as a credible and authoritative source in the health-care industry. As the digital landscape continues to evolve, prioritizing local SEO will ensure that your practice remains competitive and accessible to those in need of your services. Embracing these strategies is a step toward building a more substantial online presence and fostering trust within your local community.

To help you apply the strategies covered in this chapter, Appendix 7.1 provides a printable, nontechnical checklist you can share with your office manager or web support team.

Appendix 7.1

Here is a **printable step-by-step checklist** tailored for physicians who want to prepare their websites for AI-powered search results, such as Google's Search Generative Experience (SGE). It avoids technical jargon and can be shared with a web developer or office manager.

Physician's AI SEO Readiness Checklist

1. **Add Structured Data Tags (Schema Markup)** (time: 30–60 min; staff: web/IT or content lead)

   ☐ Go to Google's Structured Data Markup Helper.
   ☐ Paste in your website URL and choose "Local Business."
   ☐ Highlight and label key details:

      ☐ Physician name
      ☐ Clinic name
      ☐ Address and phone number
      ☐ Specialties
      ☐ Office hours
      ☐ Accepted insurance plans

   ☐ Generate JSON-LD code and embed it (via developer or WordPress plug-in like Yoast SEO/Rank Math).
   ☐ Validate using Google's Rich Results Test or Schema Markup Validator. Schema markup helps search engines understand your practice and support eligibility for rich search results.

2. **Create or Update a Patient-Friendly FAQ Section** (time: ~1 h per page; staff: content lead).

   ☐ Add a "Frequently Asked Questions" section to each service page, using a question-and-answer format.
   ☐ Write short answers to common questions, such as the following:

- ☐ What is [test or condition]?
- ☐ How do I prepare for my visit?
- ☐ What insurance do you take?
- ☐ How do I schedule or cancel an appointment?

Use FAQ schema markup to enhance visibility in search features such as rich snippets.[21]

3. **Write Content in a Question-and-Answer Style** (time: ~1 h per page; staff: content lead)

   - ☐ Use clear, conversational language.
   - ☐ Avoid medical jargon where possible.
   - ☐ Examples to include.
     - ☐ "What are the signs of diabetes?"
     - ☐ "How is high cholesterol treated?"
     - ☐ "When should I see a dermatologist?"

4. **Get Listed on Trusted Health Sites** (time: 1–2 h; staff: office manager).

   - ☐ Make sure your profile is current and accurate on the following:
     - ☐ Healthgrades, Zocdoc, Vitals
     - ☐ Local hospital or clinic websites
     - ☐ State medical society or specialty board listings
   - ☐ Confirm consistency in name, address, phone number (NAP), and alignment with your Google Business Profile and website to improve ranking and patient trust.
   - ☐ Ask local news outlets or health bloggers to link to your site if you're featured.
   - ☐ Join community directories with health sections.

5. **Use Simple Tools to Monitor Visibility** (time: 15 min weekly; staff: SEO lead or admin).

   - ☐ Sign up for Google Search Console.[22]
   - ☐ Check your site's appearance in search results.
   - ☐ Monitor changes when adding new FAQs or schema tags.

[21] **FAQ schema markup** is a special type of structured data (code) you add to a webpage that tells Google the page includes a list of frequently asked questions and their answers. When applied correctly, it allows those questions and answers to appear directly in the search results, often expanding your listing with more visible content.

[22] **Google Search Console** is a free tool from Google that allows website owners to monitor how their site appears on Google Search, troubleshoot indexing issues, view search performance data (like clicks and impressions), and receive notifications about errors or security problems.

Schema Best Practices for Physician Practices

- Use the Physician type (a subtype of LocalBusiness or MedicalOrganization) to signal provider-level detail (name, specialty, NPI, affiliations) to search engines.
- Use JSON-LD format as Google recommends it for readability and error resistance.
- Include both required properties (name, address, telephone, hours) and recommended properties (AggregateRating, specialties, insurance plans) where available.

References

1. Google/SOASTA. The need for mobile speed: how mobile latency impacts publisher revenue. 2017. Retrieved from https://www.thinkwithgoogle.com. Accessed 28 July 2025.
2. Google Search Central Page experience and search performance. 2022. Retrieved from https://developers.google.com/search/docs/appearance/page-experience. Accessed 2 Aug 2025.
3. Backlinko. Page speed study: how fast should a website load in 2023? 2023. Retrieved from https://backlinko.com/page-speed-stats. Accessed 22 July 2025.
4. Federal Communications Commission. Broadband speed guide. 2021. Retrieved from https://www.fcc.gov/research-reports/guides/broadband-speed-guide

Chapter 8
Marketing Telehealth Services

Telehealth: A Rising Trend

Telehealth expanded rapidly during the COVID-19 pandemic and remains a core component of healthcare delivery. Although the initial surge in use has leveled off, virtual care remains common across medical specialties.

By 2021, over 80% of US office-based physicians had adopted telehealth services, a sharp rise from just 16% in 2019 [1]. Physician use peaked in 2021, and recent data suggest sustained engagement, particularly in primary care and behavioral health. A 2025 study found over 70% of US physicians reported using telehealth, with live video visits the most common modality (56.5%) [2]. Payment and reimbursement were key factors enabling continued telehealth use.

On the patient side, telehealth use also surged during the pandemic. In 2021, 37% of adults reported at least one telehealth visit in the past year. This number dipped slightly to 30.1% in 2022, though broader surveys show that by 2024, 54% of Americans had used telehealth at least once. Women remain more likely than men to access care virtually.

As shown in Fig. 8.1, telehealth adoption has grown steadily, reflecting both patient demand and provider readiness.

Among Medicare beneficiaries, the proportion receiving telehealth services fell from a pandemic peak of 42% to 12.6% by the end of 2023. However, this still represents a marked increase over pre-pandemic levels, where usage was under 1% [3].

While growth has stabilized, telehealth continues to support care access, especially in hybrid models that combine virtual and in-person visits. Patients use it most often for chronic disease management, mental health, medication refills, and follow-ups. These trends show that patients expect more flexible options for receiving care.

As telehealth becomes a routine part of care, healthcare organizations must actively communicate its value to patients. This is where strategic marketing

C. Quinn, *The Digital Practice*, https://doi.org/10.1007/978-3-032-07745-5_8

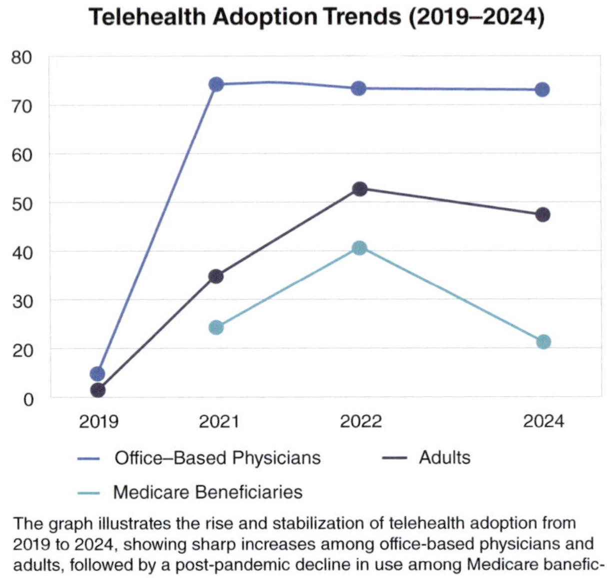

Fig. 8.1 Telehealth adoption trends. (Image Quinn/ChatGPT)

becomes essential. As adoption stabilizes, the conversation is shifting from simply offering telehealth to ensuring patients understand why and how to use it.

Why Market Telehealth Services?

Telehealth offers clear advantages for both patients and physicians. For patients, it removes barriers to care, such as travel time, mobility issues, and scheduling conflicts. It also increases the likelihood of timely follow-ups and medication adjustments, especially for those managing chronic illness. For practices, telehealth creates more efficient workflows, fills scheduling gaps, and helps retain patients who may otherwise delay or skip in-person visits. Physicians can expand their geographic reach, reduce missed appointments, and provide flexible care options without expanding physical office space.

These dual benefits include greater access for patients and increased efficiency for practices. They make telehealth a high-value service worth promoting.

Figure 8.2 summarizes the core benefits of telehealth for patients and practices, including increased access, improved efficiency, and expanded reach.

Practices that offer telehealth can attract more patients by highlighting the convenience and accessibility of telehealth. During the COVID-19 pandemic, telemedicine helped maintain safety while expanding access to care. Studies have shown that virtual visits reduce transportation costs and improve access for patients in remote or underserved areas.

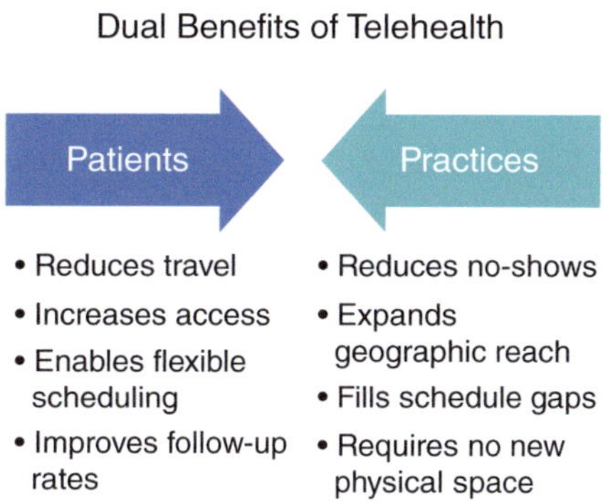

Fig. 8.2 This infographic highlights how telehealth delivers distinct advantages to both patients and healthcare practices, supporting greater access, improved efficiency, and broader reach without requiring additional physical infrastructure. (Image by Quinn/ChatGPT)

Telehealth platforms now support real-time clinical services. Multiple studies have shown no significant differences in care quality between virtual and in-person visits. A 2022 study in JAMA Network Open found that telemedicine matched or outperformed in-person care on 13 of 16 quality measures across 2.3 million primary care visits [4]. A separate review from the US Department of Health & Human Services also concluded that telehealth delivered comparable outcomes in areas such as chronic disease, behavioral health, and emergency care. Additionally, CDC survey data indicate that over 75% of primary care physicians believed they could deliver care of similar quality during telehealth visits [5]. The rapid expansion of Internet infrastructure has also made telehealth more practical for a broader range of patients.

Marketing plays a central role in promoting telehealth. Educating patients about its benefits, such as avoiding travel and receiving care from the comfort of their own homes, encourages adoption. Clear messaging also helps correct common misconceptions and increases comfort with the format. Many patients still believe that telehealth visits are impersonal, less effective, or only suitable for minor concerns. Others may assume that they need high-speed Internet or special equipment to participate. These beliefs can discourage patients from trying virtual care.

Sidebar: Common Patient Misconceptions About Telehealth

- "Telehealth is only for minor or urgent issues."
- "I won't see my regular doctor."
- "It's less secure or private than an office visit."
- "I need special equipment or a strong Internet connection."
- "It doesn't work for chronic conditions or mental health."

Physician practices can address these myths directly through clear, consistent messaging across websites, waiting rooms, appointment reminders, and community outreach materials.

To counter these assumptions, providers should use plain language and multiple channels to communicate key facts. This includes reassuring patients that telehealth can support chronic disease management, behavioral health, and follow-up care. Messages should also clarify that virtual visits utilize secure platforms and adhere to the same privacy rules as in-person care.

According to a 2022 HIMSS report, clarity and transparency in messaging improve patient trust and readiness to use telehealth services [6]. Another survey found that patients who received preparation materials before their first virtual visit were significantly more satisfied with the experience.[1]

Marketing telehealth services to existing patients reinforces their value and improves patient retention. Utilizing content marketing, blogs, and search-optimized materials can attract new patients and increase visibility. Messaging that emphasizes cost savings and ease of access appeals to patients seeking affordable and efficient care [7]. Even with clear benefits, many patients still hold misconceptions that can limit their willingness to try virtual care.

Sidebar: Checklist for Patient-Friendly Telehealth Messaging

- Use plain, reassuring language across all platforms.
- Explain how telehealth supports chronic and mental health care.
- Highlight convenience, flexibility, and time savings
- Reassure patients about privacy and provider continuity.
- Share how-to guides and short videos on how to connect.

Effective delivery methods include website banners, FAQ pages, short explainer videos, email campaigns, and in-office signage. Practices can also send text reminders with links to setup guides or provider bios to help reduce anxiety before a visit.

Effective marketing must promote telehealth's strengths while addressing concerns about quality and security. Once a practice commits to promoting telehealth, the next step is to identify which benefits matter most to patients. See Appendix 8.1 for ready-to-use marketing templates that support these efforts. Reassurance helps build trust and makes patients more likely to try virtual care.

[1]American Medical Association. (2023). Telehealth Implementation Playbook. AMA. This resource details how providing preparatory materials (appointment reminders, technology instructions, etc.) raises confidence and enhances satisfaction among telehealth users.

Promoting Telehealth Benefits

Marketing should emphasize how telehealth improves convenience, saves time, and reduces costs. Patients often value the flexibility of virtual visits, especially when juggling work, childcare, or long travel distances.

Telehealth is associated with a decreased rate of missed appointments or "no shows." A 2021–2023 retrospective cohort study at the University of South Florida compared appointment completion rates for telemedicine and in-person visits. The results showed that patients were more likely to complete scheduled telemedicine visits, suggesting that virtual care can reduce cancellations and no-shows while improving access and efficiency [8]. Figure 8.3 compares appointment completion rates for telemedicine versus in-person visits, highlighting the higher rates achieved with telemedicine.

Alanazi et al. reported that telemedicine improved satisfaction and reduced travel demands [9]. A review by Kruse et al. found that most patients associated telehealth with higher satisfaction, though opinions on efficiency and clinical outcomes varied [10]. The 2022 US Telehealth Satisfaction Study found that 94% of patients who used telehealth reported high satisfaction levels.[2]

Together, these findings show that virtual care improves engagement and convenience. Beyond general quality comparisons, some populations may benefit even more

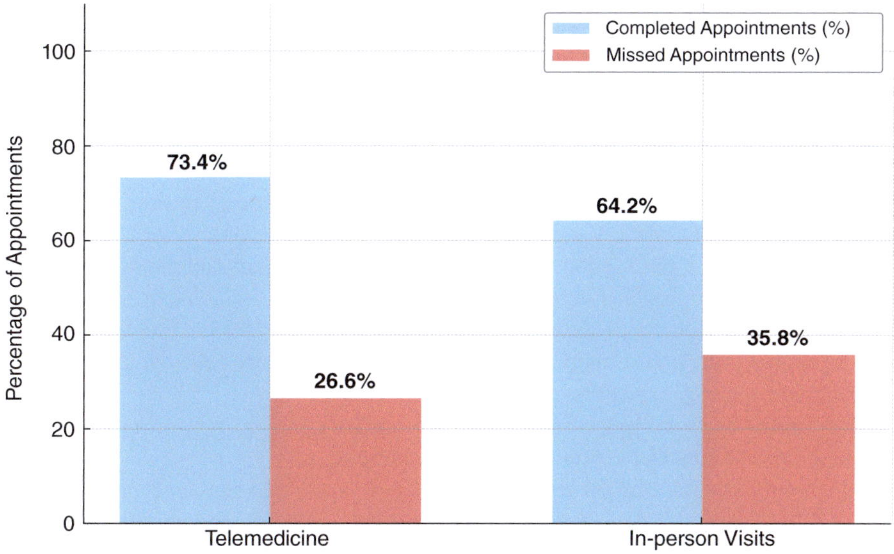

Fig. 8.3 Telemedicine achieves higher appointment completion rates compared to in-person visits (University of South Florida, 2021–2023). (Image by Quinn/ChatGPT)

[2] J.D. Power. (2022). 2022 U.S. Telehealth Satisfaction Study. J.D. Power. According to the 2022 J.D. Power Telehealth Satisfaction Study, 94% of patients who used telehealth services reported high levels of satisfaction, reflecting strong acceptance of virtual care.

from virtual care. For example, patients managing chronic conditions or receiving mental health services often report improved access, fewer missed appointments, and greater satisfaction with telehealth than with in-person care. For patients managing chronic disease or seeking mental health care, virtual visits often match or exceed in-person outcomes. Providers who highlight these benefits will appeal to patients looking for accessible, flexible care. While these benefits are compelling, lingering concerns about privacy, rapport, and technology can still prevent patients from embracing telehealth.

Marketing content should explain how telemedicine enables patients to avoid long commutes and reduce time off work. Virtual visits eliminate the need for childcare or transportation, and they also minimize waiting room delays. These points help make the value of telehealth more concrete.

A relevant study published in the *Annals of Internal Medicine* found that patients with heart failure who received early outpatient follow-up within 14 days after discharge via telemedicine had significantly lower 30-day readmission rates compared to those who received no follow-up. The study showed a 15.0% readmission rate for telemedicine visits versus 23.1% for no early follow-up, and this reduction was comparable to in-person visits (14.0% readmission) [11].

These results support telehealth as a tool for improving care delivery and outcomes. Even with growing adoption and clear benefits, some patients hesitate to use virtual care due to lingering doubts or misunderstandings. While highlighting the advantages of telehealth is important, practices also need to address common hesitations that may prevent patients from trying virtual care.

Addressing Patient Concerns About Virtual Visits

Trust is a prerequisite for adoption, and many patients remain hesitant. They are often worried about privacy, rapport with providers, or the quality of virtual care. This concern is not unfounded. Studies have found that many telehealth apps and platforms lack comprehensive privacy and security policies, which can contribute to patient apprehension about virtual visits. For example, only 28% of telehealth apps had a privacy policy, and many providers report that patients refuse telehealth services due to security concerns.[3]

The myths and realities of telehealth are outlined in Fig. 8.4, providing quick reference points to address common misconceptions.

Marketing should address these concerns with facts and reassurance. Practices can highlight provider credentials, outline security protocols, and share real patient stories to demonstrate success with telehealth.

Clear instructions help reduce confusion about how telehealth works. Providers should offer simple guides for logging in, testing equipment, and preparing for virtual appointments. This guidance reduces anxiety and increases patient confidence.

[3] Rutgers Policy Lab, 2025, https://policylab.rutgers.edu/publication/growing-cybersecurity-concerns-for-telehealth-services/. Accessed 7/30/25.

Fig. 8.4 Dispelling telehealth myths. (Image Quinn/ChatGPT)

Common Myths vs. Facts About Telehealth

Myth	Fact
✗ It's only for minor issues.	Works for chronic care and mental health
✗ It's not private	HIPAA-compliant, encrypted platforms are standard
✓ I won't see my regular doctor	Most visits are with the same provider
✓ I need special devices	A smartphone or basic internet connection works fine

By explaining the process and showcasing successful outcomes, practices can build trust and encourage more patients to try virtual care. Once these concerns are addressed, practices can focus on specific tools and tactics that effectively connect with potential telehealth users. Once these concerns are acknowledged and addressed, practices can shift their focus to the tools and templates that make telehealth marketing more efficient. This starter toolkit can be adapted to match the needs of your patient population and practice size.

Sidebar: Suggested Telehealth Marketing Toolkit
To support outreach, practices can use this toolkit:

- Sample social media captions for Facebook and Instagram.
- Email announcement templates.
- Website banner text.
- FAQ entries about telehealth security, setup, and provider access.
- Patient testimonial video scripts.
- Printable step-by-step guides for logging in.

This toolkit can be adapted to suit various patient populations and practice sizes, reinforcing key messages and reducing the workload for staff developing new materials.

Creating Content That Drives Telehealth Adoption

Strong telehealth marketing depends on the quality and relevance of the content you produce. Patients respond best to information that addresses concerns, demonstrates benefits, and makes it easy to get started with virtual care.

1. **Set a Clear Goal for Each Piece**
   Decide if your content will.
   - Explain how telehealth works.
   - Reassure patients about privacy and quality.
   - Encourage appointment scheduling.
   Knowing the purpose helps you write focused, measurable messages.
2. **Match the Format to the Audience**
   Select formats based on where your patients spend time and how they consume information:
   - **Short videos** on social media or your website showing how to join a session.
   - **Patient testimonials** in newsletters or on landing pages.
   - **Provider Q&As** in blog posts or Facebook Live sessions.
   - **Step-by-step guides** with screenshots for your website's telehealth page.
   - **Articles** on topics like "Five Tips for Your First Virtual Visit" for your blog or email campaigns.
3. **Keep Messaging Simple and Patient-Friendly**
   Avoid technical jargon. Focus on convenience, accessibility, and the ability to see the same provider as in an office visit. Use clear headlines and short paragraphs.
4. **Use Visuals to Simplify the Process**
   Show; don't just tell. Include screenshots, short clips, or infographics that explain the connection process. Add captions to videos and alt text to images, and obtain written consent before using patient likenesses.
5. **Post Consistently**
   Create a schedule patients can expect. For example.
   - Website: Update monthly with a new resource.
   - Social media: Post once or twice a week.
   - Email: Include a telehealth reminder in every newsletter.
6. **Optimize and Measure Results**
   Use location-specific keywords such as "virtual primary care in [City]." Track metrics like video views, click-through rates, and appointment bookings to refine future content.

Consistent, clear, and accessible content builds trust in telehealth. For help organizing your approach, see Appendix 8.2 Telehealth Content Planning Worksheet.

Marketing Strategies

Once your message is clear, the next step is selecting the right delivery channels. Here's where to begin. Healthcare providers can use multiple channels to promote telehealth:

- Social media: Post testimonials, answer common questions, and explain how telehealth works. Platforms like Facebook, Instagram, and YouTube reach a wide range of patients.
- Content marketing: Use blog posts, infographics, and videos to share health tips and show how telehealth supports care. Include SEO keywords to boost visibility.
- Email campaigns: Announce telehealth availability, send appointment reminders, and follow up after visits. Personalized messages help retain patients.
- Paid advertising: Use Google Ads and social media targeting to reach patients based on location, interests, or health needs.

Each tactic builds awareness and brings patients into the digital care experience. Among the most persuasive tools available, patient stories stand out for their ability to build trust and reduce uncertainty. Testimonials support outreach by persuading patients who may be uncertain about their care. Using multiple channels improves visibility and encourages adoption. Among the many marketing tactics available, patient testimonials are especially effective at turning interest into action.

Sidebar: Where to Share Patient Testimonials

To maximize their impact, share telehealth testimonials through the following:

- Practice website homepage or telehealth landing page.
- Waiting room videos or digital screens.
- Social media channels like Facebook or YouTube.
- Email newsletters or follow-up messages.
- Blog posts or case studies featuring real patient experiences.

Make sure to obtain written consent for all testimonials.

Stories should be short, authentic, and permission-approved. Highlight positive outcomes and mention any initial hesitations that were overcome.

Use of Testimonials

Real stories from satisfied patients help others understand the value of telehealth. Testimonials and case studies show how virtual care works in practice and reduce doubt among new users.

Stories often resonate more than statistics. When someone hears about a peer who saved time, avoided a long commute, or received quick care from home, they're more likely to try it themselves. Sharing stories in emails, blogs, and videos strengthens trust.

These stories also address concerns. A testimonial that mentions initial uncertainty followed by a positive outcome helps new patients feel confident about booking a virtual visit. To ensure these positive stories resonate with all patients, practices must also address the technical barriers that can limit participation.

The following example illustrates an ideal telehealth testimonial. It addresses an initial concern, describes the patient's positive experience, highlights specific benefits such as convenience and continuity of care, and ends with a personal endorsement that builds trust.

> I was hesitant to try telehealth because I wasn't sure it would feel personal. But after my first virtual visit, I was surprised at how easy it was. I saw my regular doctor, we discussed my treatment plan, and I didn't have to take time off work or arrange childcare. It felt just like being in the office, only more convenient.—Maria S., patient.

Sidebar: Supporting Patients with Limited Tech Access
Practices can improve telehealth adoption by removing digital barriers:

- Offer audio-only visits for patients without Internet.
- Use platforms that work on older phones or tablets.
- Provide loaner devices for high-need patients.
- Partner with community groups for training and tech support.
- Develop one-page guides in multiple languages with large fonts.

Inclusive design demonstrates to patients that telehealth is accessible to everyone. Highlighting these supports in marketing materials reassures them that help is available and reduces hesitation, making outreach more effective.

Address Technological Accessibility

Some patients lack reliable devices, Internet access, or the necessary technical skills. Marketing should demonstrate how the practice supports these users. Providing simple guides, offering training sessions, and offering phone support can make a significant difference.

Practices can also use telehealth platforms that work on older devices or slower connections. For example, audio-only visits and platforms optimized for low-bandwidth environments have helped reduce failed connections by over 70% in rural communities [12]. In some cases, loaner devices or community partnerships can expand access further. One community-academic program paired older adults with trained volunteers who provided tablets, one-on-one training, and ongoing support [13]. This intervention significantly increased participants' confidence in using telehealth and improved follow-through with appointments. Marketing materials that include these options show that the practice welcomes all patients, regardless of their tech readiness.

By helping patients overcome technical hurdles, providers increase satisfaction and extend the reach of virtual care.

Conclusion

Marketing telehealth allows practices to grow, retain patients, and meet rising expectations for convenience and flexibility. For patients, it offers a path to more accessible, convenient, and timely care. For providers, it improves appointment efficiency, expands geographic reach, and supports ongoing engagement. The clinical evidence confirms its effectiveness, and patient satisfaction data support its appeal.

To be successful, marketing must do more than describe features; it must address concerns, highlight real outcomes, and guide patients through the process. Strong campaigns rely on plain language, storytelling, and a mix of platforms tailored to the audience.

As virtual care becomes a permanent part of healthcare delivery, physicians who invest in thoughtful marketing will be better positioned to meet the needs of their patients and compete in a digital-first environment. Telehealth is no longer experimental. Patients expect it. Providers must communicate it clearly, confidently, and consistently. Marketing telehealth isn't about hype; it's about helping patients understand that care is now *only one click away*.

Appendixes

Appendix 8.1: Sample Telehealth Marketing Messages

These sample messages can help physician practices launch or promote their telehealth services quickly and consistently across different channels.

Email Message: Announcement of Telehealth Availability
Subject: Telehealth Appointments Now Available at [Your Clinic Name].

Dear [Patient First Name],

We're making it easier for you to get the care you need—no commute required. You can now schedule a telehealth visit with your provider at [Clinic Name]. Whether you need a follow-up, prescription renewal, or help managing a chronic condition, we're here for you—virtually.

Book your virtual appointment today: [Insert Link]

Sincerely,

[Your Clinic Name].

Social Media Caption: Facebook/Instagram
No waiting room. No traffic. No problem. You can now visit with your doctor by video from the comfort of your home. Call us or click to book a telehealth visit today. [Phone Number] [Link].

Website Banner Text
Now offering virtual visits for primary care, chronic condition management, mental health, and more. Schedule your telehealth appointment today!

FAQ Snippet for Website
Q: Is telehealth secure?
A: Yes. We use encrypted, HIPAA-compliant platforms to protect your health information during every virtual visit.
Q: What do I need for a telehealth appointment?
A: A smartphone, tablet, or computer with Internet access. No downloads or special equipment required.

Appendix 8.2: Telehealth Content Planning Worksheet

Use this worksheet to brainstorm, organize, and track telehealth marketing materials and outreach campaigns.

1. **Identify Your Audience Segments**
   - New patients
   - Existing patients
   - Seniors
   - Caregivers
   - Patients with chronic conditions
   - Mental health patients
2. **Choose Your Messaging Focus**
   - Convenience and time savings
   - Privacy and security
   - Access to your regular provider
   - Chronic care and mental health support
   - Simplicity of use
3. **Select Communication Channels**
   - Website banner or landing page
   - Email campaigns
   - Social media posts
   - SMS/text reminder
   - Waiting room flyers.

4. **Build a Content Calendar**

Week	Topic	Channel	Notes
1	Telehealth basics	Website, email	Introduce service availability
2	How-to video	Social media	Link to short video walkthrough
3	Patient story/testimonial	Blog, email	Highlight satisfaction and ease
4	Common questions answered	FAQ page	Clarify misconceptions

5. **Assign Responsibilities**
    - Content creation: [Staff Name]
    - Social media scheduling: [Staff Name]
    - Website updates: [IT contact]
    - Monitoring and analytics: [Staff Name]

These sample messages can help physician practices launch or promote their telehealth services quickly and consistently across different channels.

Email Message: Announcement of Telehealth Availability
Subject: Telehealth Appointments Now Available at [Your Clinic Name].

Dear [Patient First Name],
We're making it easier for you to get the care you need—no commute required. You can now schedule a telehealth visit with your provider at [Clinic Name]. Whether you need a follow-up, prescription renewal, or help managing a chronic condition, we're here for you—virtually.
Book your virtual appointment today: [Insert Link].
Sincerely,
[Your Clinic Name].

Social Media Caption: Facebook/Instagram
No waiting room. No traffic. No problem. You can now visit with your doctor by video from the comfort of your home. Call us or click to book a telehealth visit today. [Phone Number] [Link].

Website Banner Text
Now offering virtual visits for primary care, chronic condition management, mental health, and more. Schedule your telehealth appointment today!

FAQ Snippet for Website
Q: Is telehealth secure?
A: Yes. We use encrypted, HIPAA-compliant platforms to protect your health information during every virtual visit.
Q: What do I need for a telehealth appointment?

A: A smartphone, tablet, or computer with Internet access. No downloads or special equipment required.

Conclusion

Marketing telehealth allows practices to grow, retain patients, and meet rising expectations for convenience and flexibility. For patients, it offers a path to more accessible, convenient, and timely care. For providers, it improves appointment efficiency, expands geographic reach, and supports ongoing engagement. The clinical evidence confirms its effectiveness, and patient satisfaction data supports its appeal.

To be successful, marketing must do more than describe features—it must address concerns, highlight real outcomes, and guide patients through the process. Strong campaigns rely on plain language, storytelling, and a mix of platforms tailored to the audience.

As virtual care becomes a permanent part of healthcare delivery, physicians who invest in thoughtful marketing will be better positioned to meet the needs of their patients and compete in a digital-first environment. Telehealth is no longer experimental. Patients expect it. Providers must communicate it clearly, confidently, and consistently.

References

1. National Center for Health Statistics. Use of telemedicine among adults: United States, 2021–2022. Centers for Disease Control and Prevention. 2023. https://www.cdc.gov/nchs/products/databriefs/db493.htm
2. Carder DB, Jeon S, Baker MC, Hu X, Dill MJ, Chew LD. Telehealth modality usage by US Physicians in 2022. J Gen Intern Med. 2025; https://doi.org/10.1007/s11606-024-09287-4. Epub ahead of print. PMID: 39809963
3. American Hospital Association. Telehealth: a vital tool in modern healthcare delivery. 2025. https://www.aha.org/. Accessed 30 Jul 2025.
4. Powell RE, Henstenburg JM, Cooper G, Hollander JE, Rising KL. Patient perceptions of telehealth primary care video visits. JAMA Netw Open. 2022;5(9):e2233479. https://jamanetwork.com/journals/jamanetworkopen/fullarticle/2796668
5. U.S. Department of Health and Human Services. Telehealth and quality of care: research recap. 2024. https://telehealth.hhs.gov/documents/ResearchRecap-Telehealth_and_Quality_of_Care_09-30-24.pdf
6. Change Healthcare & HIMSS. Consumers weigh convenience and trust in telehealth decisions. 2020. https://gkc.himss.org/news/consumers-weigh-convenience-and-trust-telehealth-decisions-survey-finds
7. Medical Group Management Association. Best practices for promoting telehealth to existing patients. MGMA. 2023. https://www.mgma.com/mgma-stats/telehealth-utilization-and-patient-demand-in-2023-best-guesses-and-best-practices. Accessed 30 Jul 2025.
8. Cummins MR, Tsalatsanis A, Chaphalkar C, Ivanova J, Ong T, Soni H, Barrera JF, Wilczewski H, Welch BM, Bunnell BE. Telemedicine appointments are more likely to be completed than in-person healthcare appointments: a retrospective cohort study. JAMIA Open. 2024;7(3):ooae059. https://doi.org/10.1093/jamiaopen/ooae059. PMID: 39006216; PMCID: PMC11245742
9. Alanazi SA, Alhaqbani MH, Alsayed AF, et al. Impact of telemedicine on patient satisfaction and traveling time: a systematic review and meta-analysis. J Med Syst. 2022;46(4):47. https://doi.org/10.1007/s10916-022-01802-3.

10. Kruse CS, Krowski N, Rodriguez B, Tran L, Vela J, Brooks M. Telehealth and patient satisfaction: a systematic review and narrative analysis. BMJ Open. 2017;7(8):e016242. https://doi.org/10.1136/bmjopen-2017-016242.
11. Cheung KS, et al. Effectiveness of telemedicine visits in reducing 30-day readmissions among patients with heart failure: a retrospective cohort study. Ann Intern Med. 2022; https://doi.org/10.7326/M21-0885. Advance online publication
12. Weigel G, Ramaswamy A, Sobel L. Opportunities and barriers for telemedicine in the U.S. during the COVID-19 emergency and beyond. Kaiser Family Foundation. 2021. https://www.kff.org/womens-health-policy/issue-brief/opportunities-and-barriers-for-telemedicine-in-the-u-s-during-the-covid-19-emergency-and-beyond/
13. Lam K, Lu AD, Shi Y, Covinsky KE. Assessing telemedicine unreadiness among older adults in the United States during the COVID-19 pandemic. JAMA Intern Med. 2020;180(10):1389–91. https://doi.org/10.1001/jamainternmed.2020.2671.

Chapter 9
Measuring ROI in Digital Marketing for Healthcare Providers

Introduction to ROI in Digital Marketing

In healthcare marketing, return on investment (ROI) measures whether digital campaigns generate measurable value. ROI compares revenue from new patients, appointments, or service packages against the cost of marketing. The formula is as follows:

$$\text{ROI} = \big(\big(\text{Revenue} - \text{Cost}\big) \div \text{Cost}\big) \times 100$$

For healthcare providers and practice leaders, understanding and applying ROI is essential for making informed spending decisions and defending budget allocations.

Digital marketing enables precise audience targeting, stronger patient engagement, and more efficient scheduling [1]. Tracking ROI reveals which strategies increase inquiries, appointments, or retention. These insights help reallocate resources to the most effective channels, reduce waste, and align marketing with clinical and business objectives.

This chapter focuses on applying ROI principles in healthcare-specific contexts. You will learn to measure performance, interpret data, and connect marketing activities to tangible outcomes. Examples from clinical practices illustrate how different strategies work in real settings, and appendices provide definitions, templates, and checklists for reference.

The shift from traditional marketing to digital has allowed providers to reach patients with greater precision and lower costs [2]. Channels such as websites, email campaigns, social media, and search engine optimization now offer real-time engagement data. Unlike traditional media, digital platforms track conversions, response rates, and return visits in real time, enabling faster adjustments to improve ROI.

C. Quinn, *The Digital Practice*, https://doi.org/10.1007/978-3-032-07745-5_9

Recent survey data shows that digital campaigns often outperform traditional approaches in patient acquisition and retention [3, 4]. Digital tools also allow providers to analyze patient behavior, such as time spent on specific pages or frequency of appointment requests, and adjust campaigns accordingly. This agility helps concentrate spending on the strategies with the highest return. The sidebar that follows highlights tracking mistakes to avoid while you measure results.

Sidebar 1: Common Pitfalls in ROI Tracking

Even with solid metrics, many clinics make predictable mistakes that distort results or weaken decisions. Avoid these five common pitfalls:

1. **Attributing Results to the Wrong Campaign**
   Without proper tracking, credit may go to the wrong channel. A patient who saw both a Facebook ad and an email might be counted under only one source.
2. **Relying on Vanity Metrics**
   High page views, likes, or shares can look impressive but do not confirm patient growth or revenue gains.
3. **Overlooking Long-Term Benefits**
   Digital marketing builds trust and generates referrals, but these effects may appear months later and outside standard ROI reports.
4. **Tracking Without Clear Goals**
   ROI matters only when tied to specific, measurable targets such as increasing new patient bookings or reducing no-shows.
5. **Waiting Too Long to Review Results**
   Delaying analysis risks repeating ineffective campaigns. Reviewing performance weekly or monthly enables timely adjustments.

Focusing on meaningful data and reviewing it regularly improves the accuracy of ROI calculations and strengthens campaign performance. See Appendix 9.1 for tracking tools that support better attribution.

Digital marketing continues to evolve. Practices now use AI tools such as Salesforce Health Cloud and ChatGPT to predict patient behavior, personalize messages, and automate outreach campaigns. These changes promise to further increase efficiency, engagement, and long-term marketing returns in the healthcare setting.

To effectively improve your marketing performance, you need a structured plan for setting and evaluating measurable goals. The next section shows how to set clear, measurable goals that align with clinical priorities.

Setting Up for Success: Strategic Planning

In larger healthcare organizations, marketing teams often include digital strategists, content creators, data analysts, and outreach coordinators. These professionals collaborate with clinical and administrative leaders to set SMART goals that align with patient needs and institutional priorities. Their work includes developing campaigns, analyzing performance, and ensuring outreach supports both clinical and business objectives.

Small or solo practices may not have a dedicated marketing team. In these settings, responsibilities often fall to practice managers, office administrators, or physicians. They typically rely on tools such as website templates, email platforms, and social media schedulers. Even without a full team, using the SMART framework helps keep efforts focused and measurable.

The SMART framework consists of the following:

- **Specific**: Define goals in clear language. Example: increase website traffic by 20% in 90 days or double online flu shot bookings.
- **Measurable**: Track progress with metrics such as subscriber counts, bounce rates, or social shares.
- **Achievable**: Set challenging but realistic targets. Plan for steady growth rather than expecting to triple appointment requests overnight.
- **Relevant**: Align goals with organizational priorities. If telehealth expansion is a priority, emphasize those services across all platforms.
- **Time-bound**: Set deadlines to manage workload and maintain accountability. For instance, launch a new service campaign within 6 weeks.

Using this framework encourages regular evaluation and adjustment. It also fosters collaboration across departments through shared objectives. Once goals are defined, the next step is to ensure they reflect patient priorities before campaign design begins. Strategic alignment links marketing to clinical and operational objectives. Clinical priorities may include improving patient outcomes, expanding access, or introducing new services. Marketing priorities often focus on brand awareness, service promotion, and attracting appropriate patients. Leadership teams, including administrators, department heads, and medical directors, set these priorities based on clinical demand, resources, and community needs.

Before launching any campaign, teams should understand what patients value and how they use digital tools. This patient-centered foundation improves the likelihood of meeting both marketing and clinical goals. Use Appendix 9.3 to run a prelaunch checklist and a brief post-campaign review.

Understanding Patient Preferences

Campaign planning begins with understanding what patients want and expect. Insights can be gathered through surveys, website analytics, or feedback tools. Most patients value accurate health information, easy scheduling, and access to qualified providers.

A Pew Research Center study found that 61% of adults have searched online for health information before discussing it with their physician [5]. A 2023 Phreesia survey reported that 80% of patients prefer providers who offer online scheduling [6]. Marketing content should address these preferences by delivering reliable education, highlighting clinical expertise, and improving the user experience.

A digital campaign is a coordinated set of online activities designed to promote a service or achieve a goal. These may include email outreach, search ads, and social media posts. Unlike traditional campaigns, digital efforts provide real-time feedback and precise audience targeting. Performance can be improved by adjusting message timing, testing content formats, refining audience segments, and shifting resources to high-performing channels.

Paid search supports seasonal priorities such as flu clinics, while organic posts might focus on mental health awareness. Analytics help monitor results and refine tactics over time [7, 8].

A real-world example is Logansport Memorial Hospital's weight-management program campaign. The marketing team used Facebook and Google Ads, created a dedicated landing page, and published blog posts about the program's benefits. In 6 months, they received 46 completed forms, 54 new patient leads, and 827 visits to the landing page [9]. These outcomes aligned directly with the hospital's clinical and marketing goals.

When done consistently, strategic planning becomes a repeatable process: set realistic goals, track results, adjust campaigns, and improve continuously. This approach keeps marketing focused and builds leadership confidence in ROI outcomes. Before you launch your next digital campaign, use the Campaign Strategy Checklist in Appendix 9.3 to ensure your plan is goal-driven, measurable, and aligned with clinical priorities.

Sidebar 2: Who Does What? Marketing Roles by Practice Size

Large Healthcare Systems

- **Digital strategist:** Defines online campaign direction and aligns it with organizational goals.
- **Content creator:** Produces blogs, social posts, email copy, and videos tailored to target audiences.
- **Data analyst:** Tracks KPIs, runs tests, and reports performance insights.
- **Outreach coordinator:** Manages community partnerships, events, and patient engagement efforts.

Midsized Medical Groups

- **Marketing manager:** Oversees campaign planning, coordinates with clinical staff, and manages vendors or contractors.
- **Social media lead:** Schedules content and interacts with audiences across platforms.
- **Office administrator:** Supports marketing tasks such as newsletters, patient communication, and website updates.

Small Practices or Solo Providers

- **Physician or practice manager:** Often manages all marketing functions, including setting goals, creating content, handling local SEO, and responding to online reviews.
- **External consultant:** May provide part-time or project-based help with brand development, advertising, or website management.

Regardless of practice size, clearly assigned responsibilities and consistent use of marketing tools are essential for sustained success.

With roles set, choose channels that match your goals and your patients' behavior.

Digital Marketing Channels: A Deep Dive

Digital marketing channels are online platforms and tools that healthcare organizations use to communicate with patients, promote services, and increase visibility. These include search engines, social media platforms, email, mobile apps, and practice websites. Each channel provides different opportunities to engage patients, share health information, and drive appointments [10, 11].

1. **SEO (search engine optimization)**: SEO helps healthcare organizations improve their ranking on search engines. By optimizing web content, clinics can attract more patients searching for services like "urgent care near me" or "pediatrician in Brooklyn." This builds long-term traffic and trust at no per-click cost [12]. For example, an ENT practice increased its organic website traffic by 300% after updating its site and improving its SEO architecture, demonstrating how focused optimization can drive patient visits without paid ads.[1] Optimizing for these searches increases organic traffic and appointment requests without additional ad spend.
2. **PPC (pay-per-click)**: PPC ads, such as Google Ads or sponsored Facebook posts, target specific demographics and keywords. Hospitals can use them to promote seasonal services like flu shots or new specialty clinics. This approach provides measurable, immediate results when campaigns are tracked for conver-

[1] https://insightmg.com/case-study-ent-practice-increase-organic-web-traffic/?utm_source=chatgpt.com. Accessed 8/8/25.

sions rather than impressions alone [13]. For instance, a dermatology clinic ran a Google Ads campaign with an ad spend of $885. It generated 43,880 impressions and 2199 clicks, resulting in 103 leads. Of those, 57 patients scheduled visits—a 55.3% conversion rate from leads to patient visits.[2]

3. **Social media**: Platforms like Facebook, Instagram, and LinkedIn allow providers to share health tips, promote services, and interact with the community. For example, a clinic might post educational videos or share patient testimonials to build trust and engagement [14]. Well-planned content increases engagement, which can lead to more patient inquiries and bookings. In one Facebook Ads campaign, **Prime Women's Health** in Ontario ran targeted advertising for its MonaLisa Touch laser therapy. In just 1 month, the clinic reached 6604 people on Facebook, generated 12 new leads, and earned approximately C$18,000 in revenue. With an average cost of about C$40 per lead and strong response from targeted visuals and messaging, the campaign delivered clear, measurable ROI.[3]
4. **Email marketing**: Email campaigns keep patients informed about services, preventive care reminders, and upcoming events. A practice might send monthly newsletters with wellness tips or announcements about new providers. The ROI of email marketing in healthcare remains high due to its low cost and personalization capabilities [15]. Email marketing in healthcare earns a median ROI of around 122%, more than four times higher than other marketing channels like social media, direct mail, and paid search.[4,5] Segmenting email lists by age, visit history, or condition can further improve relevance and response rates.
5. **Mobile apps**: Mobile apps extend the practice's presence to patients' smartphones. They can offer appointment scheduling, prescription refill requests, and access to lab results or telehealth visits. Some health systems develop branded apps to push alerts, promote services, or send health tips directly to users. For example, MyChart by Epic and Healow by eClinicalWorks both support direct messaging and personalized content delivery, helping improve engagement and loyalty [16, 17]. Health systems that promote app adoption often see reduced call volumes[6] and higher patient satisfaction scores.[7] In a review of interventions using mobile-based tools, 86% of studies reported improved patient engagement

[2] https://www.omnicoreagency.com/case-studies/dermed-clinic-google-ads-ppc/?utm_source=chatgpt.com. Accessed 8/8/25.

[3] https://practicetechsolutions.com/facebook-advertising-cases/?utm_source=chatgpt.com. Accessed 8/8/25.

[4] https://blog.beehiiv.com/p/email-marketing-for-doctors?utm_source=chatgpt.com. Accessed 8/8/25.

[5] https://www.emailtooltester.com/en/blog/email-marketing-roi/#:~:text=Sources-,Email%20marketing%20ROI%20overview,ROI%20of%20their%20email%20campaigns. Accessed 8/8/25.

[6] https://www.dialoghealth.com/post/reduce-patient-call-volume. Accessed 8/8/25.

[7] https://digitalcommons.library.uab.edu/cgi/viewcontent.cgi?article=4463&context=etd-collection. Accessed 8/8/25.

[18]. One mobile intervention delivered diabetes patients personalized text messages that supported self-care and increased active involvement in their treatment [19].

6. **Websites**: A clinic's website serves as the digital front door. A user-friendly site with online booking, service pages, and provider bios helps patients find information and take action [20]. User-friendly in this setting implies the site is fast-loading, mobile-responsive, and accessible to all patients. After redesigning its site for mobile responsiveness and internal linking, a plastic surgery practice saw a **71% increase in new organic visitors** and a **28% rise in contact form submissions** year over year.[8]

Using these channels in combination amplifies outreach, improves patient experience, and advances both clinical and business goals. Success depends on tailoring each platform's strategy to how patients use it. The next section explains how to optimize each channel for measurable outcomes.

Strategies for Channel-Specific ROI Optimization

Each digital marketing channel requires a tailored approach to maximize return on investment. In healthcare, these strategies must align with specific operational goals such as attracting new patients, improving appointment adherence, or expanding awareness of service lines.

Search Engine Optimization (SEO)

SEO performs best when clinics target local search terms patients already use. Practices should create content around phrases like "urgent care for children" or "geriatrician near me" to improve unpaid search visibility. Maintaining accurate Google Business listings, updating page content regularly, and applying structured data markup can improve rankings and drive appointment requests without advertising costs. Optimizing for local healthcare keywords can deliver strong results. For a hospital website, a focused SEO strategy led to a 32% increase in overall organic traffic, a 47% rise in new users, and mobile organic sessions growing from 32% to over 50%, all within 6 months of targeted optimization efforts.[9]

---

[8] https://www.bluetonemedia.com/seo-case-study-central-florida-plastic-surgery?utm_source=chatgpt.com. Accessed 8/8/25.

[9] https://medium.com/%40nomanasghar/hospital-seo-case-study-2024-990a45eff178. Accessed 8/8/25.

Pay-Per-Click Advertising (PPC)

PPC ads reach patients searching for timely services. To improve ROI, practices should filter by demographics, avoid irrelevant keywords, and test different headlines or calls to action. Campaign success should be measured by completed actions, such as appointment bookings or phone calls, not just clicks. Directing traffic to specific service pages rather than the homepage helps increase conversion. Smaller clinics often outsource campaign setup and monitoring if internal marketing support is limited. A dermatology clinic's targeted Google Ads campaign produced 103 leads in 3 months, with 57 converting to patient visits, yielding a 55% lead-to-patient conversion rate.[10]

Social Media

Facebook, Instagram, and similar platforms help practices reach their community and humanize care. Health tips, video explainers, provider spotlights, or event promotions can engage different audiences. A regular posting schedule and prompt replies to patient messages improve visibility and build trust. Social content should support larger marketing goals, such as service awareness or seasonal campaigns. Prime Women's Health in Ontario generated C$18,000 in revenue from a single monthlong Facebook Ads campaign, with an average cost of C$40 per lead.[11]

Email Marketing

Email campaigns offer a cost-effective way to sustain patient relationships. Targeted messages based on age, visit history, or clinical need improve open rates and follow-through. For example, you might send preventive care reminders or wellness tips that link to a scheduling form. You can run small A/B tests by comparing two subject lines or two calls to action. These tests help identify which messages your audience prefers. Monitoring metrics like open rates and click-throughs allows your team to adjust timing, design, and tone. See Sidebar 3 for examples.

Mobile Apps

Mobile health apps provide ongoing access to care services. These apps let patients schedule visits, request refills, send messages, and view test results. Some systems send alerts, wellness content, or personalized follow-ups directly to the patient's

[10] https://www.omnicoreagency.com/case-studies/dermed-clinic-google-ads-ppc/?utm_source=chatgpt.com. Accessed 8/8/25.

[11] https://practicetechsolutions.com/facebook-advertising-cases/?utm_source=chatgpt.com. Accessed 8/8/25.

phone. MyChart (Epic) and Healow (eClinicalWorks) are two examples used by health systems to keep patients connected outside the clinic. In a randomized controlled trial, a mobile health intervention for hypertension management led to a 50% increase in medication adherence and significant reductions in systolic blood pressure over 12 weeks [21].

Practice Website

The website serves as the central hub for digital outreach, providing clear pathways for patients to find services, check insurance, learn about providers, and book appointments. The layout, speed, and mobile responsiveness all affect whether patients stay or bounce. Tracking analytics enable practices to refine underperforming pages and add features that support navigation or trust, such as reviews or FAQs.

When each channel is managed with clear objectives and consistent evaluation, healthcare organizations see real improvements in engagement, satisfaction, and efficiency. These outcomes contribute directly to clinical growth and financial performance.

Figure 9.1 outlines the step-by-step flow of a healthcare marketing campaign from planning and execution to measurement, analysis, and ROI-driven adjustments. Using this cycle consistently allows teams to refine campaigns, reduce wasted spend, and improve measurable outcomes.

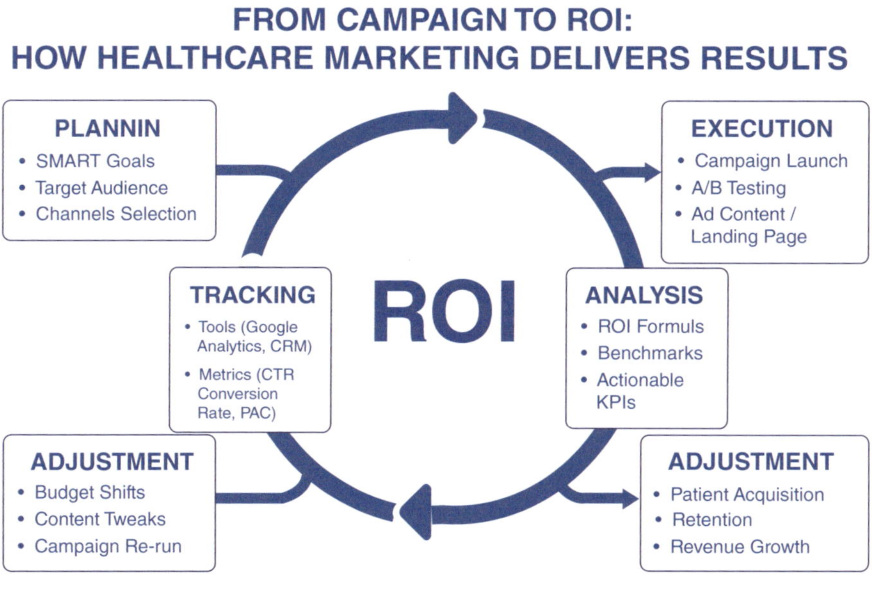

Fig. 9.1 This infographic illustrates the full cycle of a healthcare marketing campaign, from planning and execution to measurement, analysis, and ROI-driven adjustments

Sidebar 3: Simple A/B Tests for Busy Clinics

A/B testing helps clinics improve campaign performance with data-backed decisions. Even small tests can boost engagement and conversions without large budgets. Aim to run each test for at least 2–4 weeks to gather reliable results.

- **Email subject lines:** Send the same email with two subject lines (e.g., "Schedule Your Flu Shot Today" vs. "Protect Yourself This Winter"). Track open rates; a well-performing subject line can increase engagement by 10–20%.
- **Search ads:** Test two Google Ads with different headlines or calls to action. Measure click-through rates and conversion rates; even a 1% improvement in CTR can lead to a measurable increase in appointment requests.
- **Landing page copy:** Try different wording on buttons or calls to action (e.g., "Book Now" vs. "Call to Learn More"). Track the change in conversions; strong CTAs can raise conversion rates by 5–15%.
- **SMS vs. email reminders:** Send half of your appointment follow-ups by email and the other half by text. Compare confirmation rates to find the most effective channel.
- **Website imagery:** Replace a generic stock photo with a staff headshot or a local image. Track bounce rates and time on page; authentic visuals can improve engagement by up to 30%.

Use these results to refine your marketing strategies, and focus resources on the highest-performing approaches. Refer to Appendix 9.1 for tools that support test tracking.

Understanding and Utilizing Key Metrics

In healthcare digital marketing, tracking specific performance metrics helps teams measure outcomes and adjust strategies. For quick definitions and formulas, see Appendix 9.2. Key performance indicators (KPIs) are measurable values that reflect how well a marketing effort meets clinical or business goals. Common KPIs in healthcare include patient acquisition cost, click-through rates on appointment ads, conversion rates from landing pages, and engagement with educational emails [22].[12]

In contrast, actionable metrics reveal whether marketing activities produce specific, valuable results. These include the number of new patients generated by a paid search campaign, the percentage of website visitors who book an appointment, or the increase in annual wellness visits following an email reminder. Unlike vanity

[12] https://www.investopedia.com/terms/k/kpi.asp. Accessed 3/16/24.

metrics, these data points connect directly to clinical volume, revenue growth, or patient satisfaction.

By focusing on actionable KPIs, healthcare marketers can measure campaign effectiveness, identify underperforming channels, and make better decisions about where to allocate resources.[13,14] The goal is not just to generate clicks or views but to convert digital interactions into measurable outcomes that support patient care and operational efficiency [23]. Once you know what to measure, the next step is choosing the right tools to collect and interpret that data.

Some metrics look impressive but offer little strategic value. These vanity metrics, such as social media likes, shares, page views, or bounce rates, are easy to collect and often used in presentations. However, they rarely show whether a campaign improved patient acquisition or retention. A large following on Instagram, for example, does not confirm that patients are booking appointments or following care plans. These numbers can distract teams from real performance signals.

Tracking the right metrics ensures that marketing strategies deliver measurable value. These indicators reveal which campaigns work, where to invest more resources, and where to make adjustments.

- **Customer acquisition cost (CAC)**—the average cost to gain one new patient. For example, if a dental clinic spends $1200 on Facebook Ads and gains 12 new patients, the CAC is $100. In healthcare, keeping CAC under $150 is often sustainable for outpatient services.
- **Customer lifetime value (CLV)**—the total revenue a patient generates over their relationship with the practice. A family medicine clinic might calculate CLV by multiplying the average annual spend ($500) by the average retention period (8 years), resulting in a CLV of $4000.
- **Conversion rate**—the percentage of visitors who take a desired action, such as booking an appointment. If a cardiology practice's website receives 2000 monthly visitors and 80 of them schedule consultations, the conversion rate is 4%. Industry averages for healthcare websites range from 3 to 5%.
- **Return on investment (ROI)**—the net profit from a campaign relative to its cost. If a dermatology practice spends $5000 on Google Ads and earns $20,000 in booked procedures, the ROI is 300%.
- **Patient retention rate**—the percentage of patients who return for repeat visits. High retention often signals patient satisfaction and loyalty; for primary care, rates above 80% are considered strong.

By monitoring these metrics regularly, practices can adjust campaigns based on real performance data rather than assumptions. This approach maximizes ROI and supports sustainable growth. See Appendix 9.1 for the ROI Tracking Template table that records monthly cost, conversions, revenue, ROI, and patient acquisition cost.

[13] https://userpilot.com/blog/vanity-metrics-vs-actionable-metrics-saas/. Accessed 3/16/24.

[14] https://www.secoda.co/blog/how-to-know-if-your-kpi-is-a-vanity-metric. Accessed 3/16/24.

Sidebar 4 offers a quick reference to the most important marketing metrics for healthcare practices, making it easier to track performance and link campaign efforts to measurable result.

Sidebar 4: What Makes a KPI Meaningful?
Not all key performance indicators (KPIs) support better decisions. In healthcare digital marketing, a KPI is useful only if it meets three criteria:

1. **It Aligns with a Business or Clinical Goal**
   The metric should reflect progress toward an outcome such as reducing no-show rates or increasing new patient appointments.
2. **It's Simple to Measure and Interpret**
   Use straightforward data points like booked appointments, call volume, or online form submissions, rather than abstract or vanity metrics.
3. **It Triggers a Response**
   A strong KPI helps your team decide whether to continue, adjust, or replace a marketing strategy. For example, a rising patient acquisition cost may signal the need to revise ad targeting.

If a metric doesn't lead to a decision, it's not worth tracking. Focus on data that clearly shows what to do next. Common KPI definitions appear in Appendix 9.2, and examples of actionable metrics are included in the case studies later in this chapter.

Advanced Tools and Technologies

Selecting the right technology helps healthcare organizations track results, target audiences effectively, and streamline marketing operations. The following categories illustrate how these tools might be used in clinical settings. The scenarios are hypothetical and intended to show potential applications and outcomes.

- **Analytics platforms:** Tools like Google Analytics 4 track website visits, traffic sources, and user behavior. In a hypothetical example, a hospital could identify that most online appointment requests come from mobile devices, prompting a mobile-first redesign that improves conversions.
- **Customer relationship management (CRM) systems:** Healthcare-specific CRMs like Salesforce Health Cloud centralize patient communications, reminders, and referral tracking. A specialty practice might use automated recall campaigns to improve repeat appointment rates.
- **Marketing automation:** Platforms such as HubSpot or Marketo automate email campaigns, nurture patient leads, and manage social posts. An urgent care network might see higher seasonal vaccination bookings after setting up automated reminders.

- **Social listening tools:** Platforms like Hootsuite Insights or Brandwatch monitor patient reviews and social mentions. A dermatology clinic could use this data to identify emerging interest in a treatment and adjust its marketing accordingly.
- **A/B testing software:** Tools like Optimizely or Google Optimize compare versions of ads, landing pages, or emails. A physical therapy group might use testing to improve an appointment form layout and increase completions.

These scenarios demonstrate how advanced marketing tools can help healthcare organizations connect data, automate tasks, and focus resources on the initiatives most likely to improve ROI. The following section explains the core tracking systems that feed these tools with reliable data.

Tracking Systems

Tracking systems are software tools or online services that monitor and record how patients and potential patients interact with your digital marketing. They track data points like which pages people visit on your website, how long they stay, which buttons they click, how they found your site (Google search, Facebook ad, email link), and whether they schedule an appointment or complete a form. These systems are usually web-based platforms like Google Analytics, Meta Ads Manager, or call tracking services. You access them through a browser on your computer, often with a log-in provided when the system is set up.

To use a tracking system, you (or your web team) place a small code on your website or landing pages. This code collects anonymous visitor data and sends it to your tracking dashboard. You can then view reports showing how well your marketing is working. For example, you can see which ad campaigns are driving calls, which web pages are most visited, and where visitors drop off before converting. This helps you adjust your marketing budget and messaging to focus on what gets results, like running more of the ads that lead to booked visits and fixing parts of the website that lose visitors.

The benefit is simple: instead of guessing what's working, you have actual data guiding your marketing decisions.

To deploy a tracking system, practices must first define clear marketing goals (e.g., increasing appointment requests from paid ads or improving engagement with newsletters). Then, they select tracking tools that align with those goals, integrate them into their digital platforms, and train staff on usage. As data flows in, the team evaluates campaign performance, identifies trends, and adjusts outreach efforts based on real-time or historical insights. Over time, this process improves marketing efficiency, lowers acquisition costs, and strengthens patient relationships. The examples below are illustrative and intended to show potential applications.

- **Call tracking**—assigns unique phone numbers to different marketing campaigns so practices can see which ads or platforms drive the most calls. For example, a

clinic might find that its Google Ads generate twice as many calls as its social media campaigns.
- **Form tracking**—monitors which web forms are completed and from which traffic sources. This could reveal that patients who arrive from a blog post about preventive care are more likely to schedule a wellness visit than those from a general homepage link.
- **UTM parameters**—adding unique tracking codes to URLs allows marketers to attribute online actions to specific campaigns, such as differentiating between email, Facebook, and paid search visitors.
- **CRM integration**—linking call tracking, form data, and appointment systems into a CRM provides a complete view of the patient journey from first touchpoint to visit.

By using these tools, practices gain a clearer understanding of how each marketing channel contributes to results, enabling better allocation of time and budget. With accurate data flowing, AI can surface patterns and guide real-time adjustments.

AI for Data Analysis and Pattern Recognition

This strategy uses artificial intelligence algorithms to scan large datasets for trends, correlations, and behavioral insights. AI can uncover patterns in how patients respond to marketing messages or interact with websites, patterns that often go unnoticed by human analysts.

For example, a machine learning model might detect that patients who watch 30-second educational videos are twice as likely to schedule an appointment. Or, it might identify that evening Facebook posts generate more engagement than morning ones. This insight enables more precise targeting, content scheduling, and ad spend allocation. For physician-marketers, the benefit lies in eliminating guesswork and focusing the budget on tactics with proven effectiveness.

Example A dermatology clinic uses AI to analyze form submissions and appointment bookings. It finds that patients who enter the site through Google search convert at a higher rate than those from Instagram. Based on this, the clinic increases its SEO investment and adjusts its content strategy. See Appendix 9.3 for a list of AI analytics platforms and comparison features.

AI for Predictive Analytics

Predictive analytics uses past behavior to forecast future outcomes. In healthcare marketing, it can help predict which patients are likely to respond to reminders, return for follow-up care, or inquire about new services. AI models build these predictions by analyzing data such as click behavior, visit history, demographics, and communication preferences.

For physicians, predictive tools support personalized outreach and smarter campaign timing. Instead of sending generic reminders, the system can flag when a patient is most likely to book an appointment and send a tailored message accordingly.

Example A pediatric practice reviews engagement data from back-to-school immunization campaigns. AI predicts that parents open emails most often between 7 and 9 a.m. on weekdays. The office shifts its outreach to match that timing, increasing appointments during August and September.

AI for Personalization at Scale

This strategy involves tailoring messages, offers, and educational content to individual patient profiles. AI-driven personalization uses real-time data (e.g., website behavior, diagnosis history, or prior visits) to send content that feels highly relevant to the recipient.

For example, an AI system can send targeted reminders to diabetic patients about foot care, flu shots, or new medication options. This builds trust and engagement while reinforcing clinical goals. The benefit for physicians is improved adherence, stronger patient satisfaction, and greater loyalty, along with higher response rates from marketing efforts.

Example A family medicine clinic integrates AI with its email system. When a patient clicks on an article about high blood pressure, they receive a follow-up email with information about local screening services and lifestyle tips, increasing patient education and appointment rates.

AI for Real-Time Campaign Optimization

This strategy allows campaigns to adjust automatically based on real-time performance data. If an ad underperforms, AI can pause it and redirect the budget to a more effective message. If a particular headline generates more clicks, the system can update all active campaigns to use that version.

For physician practices, this approach minimizes waste and maximizes returns without requiring constant oversight. It helps small marketing teams maintain efficient and responsive campaigns.

Example A women's health center launches a digital ad campaign promoting same-day telehealth visits. AI detects that ads mentioning "no wait times" perform best. Within hours, the system updates all ad creatives to use this phrase, increasing click-through rates and patient bookings. With these AI tools in place, practices can not only improve day-to-day results but also measure long-term financial impact more accurately. The next section translates those results into ROI and lifetime value.

Calculating ROI: Advanced Techniques

Understanding the return on investment (ROI) from healthcare marketing helps practices measure effectiveness and guide future spending. ROI reflects how much revenue your marketing efforts generate compared to what you spent. To begin, add up the total cost of the campaign. This includes expenses like ad spend, content creation, software subscriptions, and staff time. Then calculate the total revenue that came directly from the campaign, such as income from new patient visits, procedures, or service packages.

The basic formula is as follows:

$$\text{ROI} = \left(\left(\text{Total Revenue} - \text{Total Cost}\right) \div \text{Total Cost}\right) \times 100$$

In healthcare, basic ROI only tells part of the story. Many campaigns have delayed or cumulative effects, especially when patients return for future services or refer others. To capture these long-term benefits, marketers use additional metrics.

Patient lifetime value (PLV) estimates the total revenue one patient will bring over the course of their relationship with your practice.

Cost per acquisition (CPA) measures how much you spent to acquire each new patient through a specific campaign.

For example, suppose a clinic runs a digital campaign costing $10,000. It attracts 100 new patients. If the estimated PLV per patient is $1000, then

- Total revenue = 100 patients × $1000 = $100,000
- ROI = (($100,000 – $10,000) ÷ $10,000) × 100 = **900%**

This result gives a more accurate picture of long-term value than looking at first-visit revenue alone.

Tools like Google Analytics, CRM dashboards, and revenue attribution platforms can support ROI tracking by tying campaign data to patient actions. Some practices also apply predictive analytics or machine learning tools to forecast future revenue and fine-tune campaign budgets based on patterns in the data.

By using ROI calculations that include both direct and lifetime revenue, physicians and marketers can make more informed choices about where to invest and which strategies to scale.

To apply these calculations in your own campaigns, use the ROI Tracking Template in Appendix 9.1 to log costs, conversions, and patient acquisition metrics over time. Accurate ROI work also depends on lawful and ethical use of data, which the next section addresses.

Navigating Challenges in ROI Measurement

Healthcare marketers face unique obstacles when measuring return on investment (ROI), due to strict privacy laws, regulatory rules, and ethical responsibilities.

Privacy Regulations , especially HIPAA in the United States, limit how marketers use patient data. Providers cannot track or personalize campaigns using patient records unless patients give explicit consent. This restriction makes it harder to attribute conversions or engagement directly to specific digital marketing efforts.

Compliance introduces additional barriers. Marketing content must meet legal standards for medical advertising. For example, if a campaign promotes a medical service, it must include evidence-based claims and disclose potential side effects. These rules shape both the content and strategy of campaigns.

Ethical Responsibility requires marketers to protect patients from misleading messages or manipulative tactics. They must avoid content that targets vulnerable individuals or overstates medical benefits. Instead, campaigns should inform, educate, and guide patients toward appropriate care options.

To manage these challenges, healthcare organizations need privacy-conscious analytics strategies. Aggregated metrics and anonymized data allow teams to assess campaign performance without exposing individual patient information. For example, marketers can analyze trends in portal log-ins or appointment requests without tying them to identifiable records.

In parallel, transparent communication builds trust. Teams should explain how they handle user data and ensure that every message reflects medical accuracy and clear intent. Focusing on patient education, rather than persuasion, helps organizations meet marketing goals while maintaining credibility.

Finally, privacy and ethics constraints also affect how teams calculate ROI. Incomplete or anonymized data can reduce the precision of key performance metrics. Marketers must recognize these limits and build evaluation frameworks that emphasize trust and compliance alongside financial results.

By combining ethical communication, compliant content, and responsible data use, healthcare marketers can balance regulatory demands with effective performance tracking. Even with privacy constraints, marketers can use responsible analytics to improve campaign effectiveness and guide future planning. The next section outlines a stepwise process for turning raw data into decisions.

Analyzing and Interpreting Data

Healthcare marketing teams rely on a structured process to evaluate campaign performance and improve future strategies. This process begins with gathering data from key sources, including website analytics, social media activity, email click-through rates, and patient inquiries. Teams review this information to track which campaigns generate appointments, attract new patients, or promote specific services.

After data collection, marketers clean and standardize the datasets. This step removes errors, unifies formats, and ensures consistency across platforms. Clean data provides a solid foundation for accurate analysis.

Marketers then apply tools like **regression analysis**, a method that identifies the strength and direction of relationships between variables. For example, it can show how changes in ad frequency affect appointment bookings. They also use **cohort analysis**, a technique that aggregates groups of users based on shared characteristics or behaviors over time, such as new patients who joined in the same month or those referred through a particular channel. This allows teams to compare how different segments respond to campaigns.

To forecast results, marketing teams use predictive modeling. This technique analyzes historical data to estimate future outcomes. A team might use predictive models to determine which message format leads to the highest appointment rates or which channel drives long-term patient engagement. When done well, this analysis improves both strategy and budget allocation.

Artificial intelligence expands these capabilities. AI platforms process large volumes of data quickly and identify trends that might go unnoticed. For example, an AI system could detect that older adults respond better to appointment reminders sent in the morning, while younger patients engage more with evening texts. These insights allow campaigns to target each group more effectively.

AI also enables real-time optimization. Tools like Salesforce Einstein or Adobe Sensei monitor ongoing campaign performance and adjust content, placement, or timing without human intervention. This dynamic feedback loop improves efficiency and increases ROI.

Finally, AI supports personalization at scale. By analyzing patient behavior, AI tools can send customized reminders, recommend relevant services, or follow up with educational content. A patient managing asthma might receive a tailored message with seasonal health tips and a link to book a visit. These targeted messages increase engagement and improve outcomes.

When healthcare marketers combine traditional analytics with AI-driven tools, they gain a deeper understanding of what works, why it works, and how to refine it. This approach supports smarter campaigns, better patient communication, and a stronger return on investment. These principles come to life in real-world case studies that show how analytics improve outreach, engagement, and ROI.

Case Studies: Using Analytics to Improve Healthcare Marketing

These case studies show how healthcare organizations use analytics to improve marketing campaigns, strengthen patient engagement, and increase ROI.

Case Study 1: Social Media Campaign to Increase Appointment Bookings

A healthcare provider launched a social media campaign to increase patient engagement and boost appointment volume. The team posted health tips, patient testimonials, and interactive content to connect with the community. They tracked likes, shares, click-throughs, and appointment bookings that followed each post.

The team used **regression analysis** to evaluate how social media engagement influenced appointment rates. They found a direct, positive link: when engagement rose, appointments increased. They then applied **predictive modeling** to identify the best times and days to post. These insights helped them revise their posting schedule and tailor their content strategy.

As a result, the provider increased both online engagement and appointment volume. Their improved marketing strategy helped more patients access care while delivering a strong return on investment.

Case Study 2: Optimizing Email Campaigns with Segmentation and Forecasting

A physician group created an email campaign to encourage patients to schedule appointments. The team used **cohort analysis** to segment recipients based on demographics and past visits. Each group received emails tailored to its preferences and behavior.

This approach improved open rates and led to more scheduled appointments. The team then tested subject lines and send times using **predictive analytics**, forecasting which combinations performed best and updating the campaign accordingly.

With this strategy, the group reduced email waste, improved communication, and increased revenue from scheduled visits.

Case Study 3: Improving Patient Portal Use with Experience-Based Analytics

A hospital system promoted its patient portal through email, social media, and its website. Website traffic increased, but few patients registered or logged in.

The team utilized **predictive analytics** to analyze user behavior, finding that patients encountered difficulties with a confusing sign-up process and suboptimal portal design. These obstacles deterred them from using the system.

The hospital responded by redesigning the portal, simplifying the layout, reducing the steps in the registration process, and highlighting popular features such as lab results and appointment scheduling.

After the redesign, more patients registered and utilized the portal. Phone call volume decreased, administrative tasks were reduced, and patient satisfaction improved. The campaign delivered measurable value by making digital tools easier for patients to use. As marketing becomes more sophisticated, emerging technologies promise even greater precision and personalization. The following section summarizes near-term trends that will shape measurement and ROI.

Future Trends and Innovations

Healthcare digital marketing continues to evolve through emerging technologies that improve how teams measure and improve return on investment (ROI).

Artificial intelligence (AI) and machine learning now play a central role in campaign optimization. These tools process large datasets quickly, identify engagement patterns, and deliver insights that help marketers refine messages and timing. For example, predictive algorithms can anticipate consumer behavior and recommend when and how to launch content for stronger results.

Digital transformation across the healthcare sector also supports ROI improvements. Electronic health records (EHRs) enable more coordinated care by reducing duplication and improving patient tracking. Wearable devices collect real-time health data that supports remote monitoring and intervention. These technologies help clinicians manage care more efficiently while giving marketers new, anonymized data streams for evaluating outreach effectiveness.

On the marketing side, digital transformation allows providers to connect with patients through mobile apps, social media, and personalized email campaigns. These tools support engagement by sharing educational content, reminders, or health tips directly with users. Marketers can then measure how these interactions influence satisfaction, appointment volume, or adherence.

Voice search also shapes how patients find care. As more people use smart speakers and mobile assistants to look up symptoms or providers, marketers must adjust their content to match conversational queries and local search terms. Optimizing for voice improves visibility and drives inbound traffic.

Looking ahead, healthcare marketers will need to adopt more personalized, privacy-conscious strategies. Data-driven personalization can improve performance, but marketers must stay within the boundaries of HIPAA and consumer privacy laws. Technologies that enable segmentation, targeting, and ROI tracking will need to incorporate ethical and legal safeguards.

To succeed, marketers should invest in platforms that combine data analytics, compliance features, and omnichannel delivery. Those who adapt quickly will gain an advantage in both patient engagement and ROI measurement.

Conclusion

Measuring ROI in healthcare marketing requires a combination of clear objectives, accurate tracking, and the right analytical tools. Metrics such as CAC, CLV, conversion rate, and retention rate provide the foundation for evaluating campaign performance. Tracking systems connect specific marketing efforts to measurable outcomes, while advanced technologies and AI offer opportunities to refine targeting, automate analysis, and predict patient behavior.

When these elements work together, marketing decisions shift from guesswork to evidence-based strategy. The result is more efficient use of resources, higher patient engagement, and stronger financial returns. As healthcare organizations continue to adopt digital tools, the ability to measure and improve ROI will remain central to sustaining growth and delivering better patient experiences.

Appendices

Appendix 9.1: ROI Tracking Template

Use this table to track and evaluate the performance of your digital marketing campaigns. Update it monthly or quarterly. Review results as part of your team's performance discussions.

Month	Campaign name	Marketing channel	Cost ($)	# of conversions	Revenue ($)	ROI (%)	PAC ($)	Goal Met?	Notes
January	Knee pain campaign	Google Ads	2000	25	5500	175%	80.00	Yes	Improved conversions after CTA update
February	Flu shot reminder	SMS/email	500	60	3000	500%	8.33	Yes	Email outperformed SMS
March	Diabetes workshop	Facebook Ads	1000	15	2250	125%	66.67	No	Consider adding retargeting ads

Definitions

- **# of conversions**: The number of users who completed a desired action (e.g., appointment booked, form submitted).
- **Revenue**: Total revenue generated from patients who converted through the campaign.
- **ROI (return on investment)**
  Formula: ((Revenue – Cost) ÷ Cost) × 100
- **PAC (patient acquisition cost)**
  Formula: Cost ÷ # of Conversions

- **Goal Met?**: Mark "Yes" or "No" based on your campaign's target for conversions or revenue.

> **Tip:** Customize this template to fit your marketing goals. Keep it in a shared folder and revisit it monthly as part of your review process.

Appendix 9.2: Glossary of Marketing Terms

This glossary defines key terms used throughout this chapter. It is intended for healthcare providers and administrators working with digital marketing teams or managing outreach campaigns.

- **A/B Testing**
  A method that compares two versions of a digital message (like an email or ad) to see which performs better.
- **Bounce Rate**
  The percentage of website visitors who leave after viewing only one page. High bounce rates may indicate irrelevant or unclear content.
- **Call to Action (CTA)**
  A phrase that directs users to take a specific step, such as "Book Now" or "Learn More."
- **Click-Through Rate (CTR)**
  The percentage of users who click on a link or ad after seeing it. A higher CTR usually means stronger engagement.
- **Conversion**
  A completed action you want from the user, such as booking an appointment, calling your office, or downloading a guide.
- **Conversion Rate**
  The percentage of users who complete a conversion out of the total who interacted with your content.
  Formula: (Conversions ÷ Total Visitors) × 100
- **Cost per Acquisition (CPA)**
  The average cost to acquire one customer or patient.
  Formula: Total Cost ÷ Number of Conversions
- **Digital Campaign**
  A set of coordinated online activities designed to promote a specific service or reach a defined goal.
- **Engagement Rate**
  A measure of how users interact with your content, including clicks, likes, comments, and shares.

- **Google Analytics**
  A website analytics tool that shows how users find and navigate your site. It helps track traffic sources, page views, and user behavior.
- **Impressions**
  The number of times your ad or content is displayed to users, whether or not they click.
- **Key Performance Indicator (KPI)**
  A specific metric that shows how well your campaign is performing relative to a goal, such as appointment bookings or patient inquiries.
- **Landing Page**
  A standalone web page designed to convert visitors. It often includes a focused message and a call to action.
- **Lead**
  A potential new patient who shows interest by completing a form, calling the clinic, or signing up for a newsletter.
- **PAC (Patient Acquisition Cost)**
  A healthcare-specific version of CPA.
  Formula: Cost ÷ Number of New Patients Acquired
- **Predictive Analytics**
  A method that uses past data and machine learning to estimate which patients are likely to take action, such as booking a visit.
- **Retargeting**
  A digital strategy that displays ads to users who previously visited your website or interacted with your content.
- **Return on Investment (ROI)**
  A percentage showing how much revenue a campaign generated compared to its cost.
  Formula: ((Revenue – Cost) ÷ Cost) × 100
- **Search Engine Optimization (SEO)**
  A strategy to increase the visibility of your website in unpaid search engine results by improving content quality and structure.
- **Vanity Metrics**
  Data points that appear positive (e.g., likes or page views) but do not necessarily indicate business impact or patient growth.

Appendix 9.3: Campaign Strategy Checklist

This checklist helps healthcare practices plan and evaluate digital marketing campaigns. Complete it before launch to improve focus, efficiency, and ROI. Use it again after the campaign ends to review outcomes and guide future strategies.

Planning Checklist: Before Campaign Launch

Goal and Targeting

- Defined a clear, measurable campaign goal (e.g., increase sports physicals by 15%)
- Confirmed alignment with clinical or business priorities
- Identified the target audience (e.g., parents of teens, Medicare patients)
- Segmented the audience if needed (e.g., by age, condition, location, insurance)

Budget and Roles

- Set a specific budget for ads, tools, or vendor support
- Assigned team members or contractors to key roles
- Collected needed materials (photos, provider bios, clinic hours)

Channel and Content Planning

- Chose the right marketing channels (email, SMS, search ads, social media)
- Wrote messages with clear calls to action (e.g., "Book Now," "Call Today")
- Created or updated the landing page to support conversions

Timing and Testing

- Scheduled the campaign for an appropriate time window (e.g., flu season, back-to-school)
- Ran A/B tests for subject lines, ad copy, or CTAs
- Tested links, buttons, and tracking tools before launch

Privacy, Compliance, and Ethics

- Confirmed HIPAA compliance for all tools and messages.
- Avoided misleading claims and included accurate service details.
- Ensured data tracking tools use anonymized, consent-based methods.

Measurement and Benchmarks

- Selected KPIs such as CTR, conversions, ROI, or patient acquisition cost.
- Reviewed performance from similar past campaigns for comparison.
- Installed or configured analytics tools (e.g., Google Analytics, CRM dashboard).

Contingency Readiness

- Identified a backup plan if performance is low (e.g., shift budget, change content).
- Assigned team contact for urgent fixes or changes during campaign.

Post-Campaign Review

Performance Summary

- Collected results for all selected KPIs
- Calculated ROI and PAC using Appendix 9.1
- Compared campaign performance to benchmarks or previous efforts

Lessons Learned

- Identified what worked well (e.g., timing, content, channel)
- Noted areas for improvement
- Saved campaign materials, results, and notes for future planning

Using this checklist keeps your campaigns structured, patient-focused, and measurable. Save each version as part of your practice's marketing records.

References

1. Accenture. Digital health technology vision. 2021. https://www.accenture.com/us-en/insights/health/technology-vision-health.
2. American Hospital Association. Hospitals use digital strategies to reach patients. 2021. https://www.aha.org/news/headline/2021-03-22-hospitals-use-digital-strategies-reach-patients.
3. Thomas EE, Caputi P, Wilson RL. Telehealth utilization in Australia during the COVID-19 pandemic: trends and implications for future health policy. Aust Health Rev. 2020;44(5):582–5. https://doi.org/10.1071/AH20161.
4. Cardinal Digital Marketing. The state of healthcare marketing. 2022. Retrieved from Cardinal Digital Marketing survey report on healthcare marketing strategies, including digital channels for patient acquisition. arxiv.org
5. Pew Research Center. The social life of health information. 2009. https://www.pewresearch.org/internet/2009/06/11/the-social-life-of-health-information/. Accessed 1 Aug 2025.
6. Phreesia. Patient perspectives on scheduling and engagement tools. 2023. https://www.phreesia.com/insights. Accessed 1 Aug 2025.
7. Radu M, Glont M, Tigan S. Online marketing campaigns—efficiency evaluation methods. Ovidius Univ Annal Ser Eco Sci. 2017;*XVII*(2):635–40.
8. Deloitte Insights. Digital transformation in healthcare in 2020: the age of human experience. Deloitte Development LLC. 2020. https://www.deloitte.com/us/en/insights/industry/health-care/digital-transformation-in-healthcare.html. Accessed 1 Aug 2025.
9. TBH Creative. Logansport Memorial Hospital marketing case study: weight management campaign. 2023. https://www.tbhcreative.com/project/logansport-memorial-hospital-hmr-campaign. Accessed 1 Aug 2025.
10. Healthgrades. Digital marketing strategies for modern healthcare practices. 2023. Retrieved from https://www.healthgrades.com/. Accessed 8 Aug 2025.
11. Becker's Healthcare. How digital transformation is shaping the patient experience. 2022. Retrieved from https://www.beckershospitalreview.com/. Accessed 8 Aug 2025.
12. Forrester Research. The long-term value of SEO in healthcare marketing. 2023. Retrieved from https://go.forrester.com/. Accessed 8 Aug 2025.
13. Google Ads. Drive health campaigns with PPC. 2023. Retrieved from https://ads.google.com/. Accessed 8 Aug 2025.
14. Pew Research Center. Social media use in 2023. 2023. Retrieved from https://www.pewresearch.org/. Accessed 8 Aug 2025.
15. Campaign Monitor. The ROI of email marketing for healthcare providers. 2023. Retrieved from https://www.campaignmonitor.com/resources/. Accessed 8 Aug 2025.
16. Epic Systems. MyChart product overview. 2023. Retrieved from https://www.epic.com/software#PatientEngagement. Accessed 8 Aug 2025.
17. eClinicalWorks. Healow: health & online wellness. 2023. Retrieved from https://www.eclinicalworks.com/. Accessed 8 Aug 2025.
18. Sawesi S, Rashrash M, Phalakornkule K, Carpenter JS, Jones JF. The impact of information technology on patient engagement and health behavior change: a systematic review of the

literature. JMIR Med Inform. 2016;4(1):e1. https://doi.org/10.2196/medinform.4514. PMID: 26795082; PMCID: PMC4742621.

19. Yan AZ, Staab EM, Nuñez D, Zhu M, Wan W, Schaefer CT, Campbell A, Quinn MT, Baig AA. Impact of a text messaging intervention as an in-between support to diabetes group visits in federally qualified health centers: cluster randomized controlled study. JMIR Diabetes. 2024;9:e55473. https://doi.org/10.2196/55473.
20. NexHealth. Modern healthcare websites: What patients expect. 2022. Retrieved from https://www.nexhealth.com/. Accessed 8 Aug 2025.
21. Morawski K, Ghazinouri R, Krumme A, Lauffenburger JC, Lu Z, Durfee E, Oley L, Lee J, Mohta N, Haff N, Juusola JL, Choudhry NK. Association of a smartphone application with medication adherence and blood pressure control: the MedISAFE-BP randomized clinical trial. JAMA Intern Med. 2018;178(6):802–9. https://doi.org/10.1001/jamainternmed.2018.0447.
22. Mihas P. Key digital marketing metrics for Healthcare. Healthcare Marketing Review; 2023.
23. Chaffey D. Digital marketing: strategy, implementation and practice. 8th ed. Pearson Education; 2022.

Chapter 10
Leveraging Emerging Technologies in Digital Marketing

This chapter examines how artificial intelligence (AI), virtual reality (VR), augmented reality (AR), and messaging platforms such as SMS and WhatsApp transform patient engagement and healthcare marketing. These tools are already widely used to personalize outreach, support patient education, and reduce missed appointments [1, 2].

SMS appointment reminders significantly improve attendance—systematic reviews show they reduce missed appointments by 20–30% [3–5]. Yet nearly one-third of healthcare providers mistakenly believe consumer texting apps are HIPAA-compliant, even though platforms like WhatsApp lack essential features such as audit logs and business associate agreements.,[1,2] To protect patient privacy and stay compliant, healthcare organizations should use secure messaging platforms that include encryption, audit trails, and role-based access controls.[3] AI helps analyze clinical and behavioral data to automate personalized messaging and improve retention [6].

Virtual reality enables immersive patient education and therapeutic intervention. Research shows that VR improves patient satisfaction and knowledge retention, especially in rehabilitation and pain management settings [7, 8].

Augmented reality overlays digital content onto real-world environments to support clinical training, patient education, and guided procedures. Recent data show that over 70% of US medical schools now incorporate VR into their training programs; studies report a 30% improvement in procedural performance when using VR compared to traditional methods.[4] A broader review found VR consistently

[1] https://www.dialoghealth.com/post/healthcare-texting-statistics?utm_source=chatgpt.com

[2] https://www.dialoghealth.com/post/hipaa-compliant-texting?utm_source=chatgpt.com. Accessed 8/9/25.

[3] https://luxsci.com/is-whatsapp-hipaa-compliant/?utm_source=chatgpt.com. Accessed 8/9/25.

[4] https://www.globenewswire.com/news-release/2024/10/28/2970114/0/en/Virtual-Reality-VR-in-Healthcare-Market-will-Reach-USD-46-40-Billion-by-2032-Driven-by-increasing-adoption-of-immersive-technologies-in-medical-applications-Research-by-SNS-Insider.html?utm_source=chatgpt.com. Accessed 8/9/25.

C. Quinn, *The Digital Practice*, https://doi.org/10.1007/978-3-032-07745-5_10

boosts learner satisfaction, confidence, and patient safety awareness, even though fewer studies directly link to clinical outcome improvements [9, 10]. Meanwhile, the global VR healthcare market is growing—from USD 3.12 billion in 2023 to an estimated USD 46.37 billion by 2032, with a compound annual growth rate of 35%.[5]

Together, these technologies enhance how providers communicate and engage with patients. Practices that utilize them report smoother workflows, higher patient satisfaction, and measurable improvements across clinical and marketing outcomes. These immersive technologies represent only one part of the expanding toolkit for enhancing patient engagement and professional education, paving the way for other digital innovations now entering healthcare practice. Among these emerging tools, artificial intelligence stands out as the most versatile, influencing not only patient communication but also predictive care planning and marketing strategy.

Exploring the Potential of AI in Patient Engagement and Education

Artificial intelligence (AI) tools, such as chatbots, predictive models, and automated messaging systems, now support personalized patient education and clinical outreach. These technologies enhance communication by analyzing patient data to tailor content, anticipate questions, and deliver timely health information [11].

AI can also interpret large, complex datasets to identify individual patient risks. Machine learning models analyze electronic health records, claims data, and diagnostic imaging to detect early signs of chronic conditions and suggest preventive strategies [12]. This pattern recognition helps physicians prioritize high-risk patients and allocate resources more efficiently [6].

For example, predictive analytics based on historical health records can forecast potential events, such as hospital readmissions, disease progression, or patient noncompliance. These forecasts enable providers to take early action and reduce complications [13].

Case in Point In one hospital, an AI-driven decision support tool identified high-risk patients among 2460 admissions. After 6 months, readmissions fell from 11.4% to 8.1%. Adjusted for control hospitals, this represented a 25% relative reduction, with 1 readmission prevented for every 11 high-risk patients treated.[6] These data demonstrate how predictive analytics can guide targeted interventions and enhance care outcomes while reducing hospital returns.

[5] https://www.fortunebusinessinsights.com/industry-reports/virtual-reality-vr-in-healthcare-market-101679?utm_source=chatgpt.com. Accessed 8/9/25.

[6] https://www.mghihp.edu/news-and-more/opinions/data-analytics/data-decisions-leveraging-predictive-analytics-transform-hospital-readmissions?utm_source=chatgpt.com. Accessed 8/9/25.

When integrated into patient outreach and marketing platforms, AI enhances engagement, drives retention, and supports better clinical outcomes through more intelligent targeting and content delivery. However, these tools also raise concerns about patient data privacy, potential algorithmic bias, and the need for clear governance on how recommendations are generated.

When using AI-driven outreach, inform patients when a message is generated or assisted by automation, and clearly explain how their data will be used. Provide concise, plain-language disclosures that specify the purpose of outreach, the data sources involved, and any third parties with access. Maintain a record of patient consent, including the date, method, and scope of permission granted. These measures help maintain trust, comply with privacy regulations, and give patients control over their personal information. While AI excels at analyzing data and personalizing outreach, immersive technologies such as virtual and augmented reality engage patients through direct, experiential learning.

Sidebar 1: Practical AI Tools for Healthcare Marketing

Explore these real-world tools that help clinics and hospitals apply AI in everyday marketing:

- **ChatGPT (OpenAI):** Generates FAQs, blog posts, and social media captions tailored to patient concerns. Reduces writing time and improves content consistency
- **Persado:** Uses AI to analyze emotional tone and optimize email subject lines, SMS content, and ad copy for better conversion rates
- **Healthwise:** Embeds AI-powered patient education into portals and email campaigns. Personalizes learning materials based on patient demographics and conditions
- **Salesforce Health Cloud:** Applies predictive analytics to time outreach campaigns, suggests next-best actions, and segments patient populations for targeted communication

These tools help automate and personalize patient engagement while maintaining a consistent brand voice. When AI tools process or store protected health information (PHI), healthcare organizations must ensure compliance with HIPAA by executing Business Associate Agreements (BAAs) with vendors.

Virtual Reality and Augmented Reality in Patient Education

Virtual reality (VR) and augmented reality (AR) technologies provide practical, clinically relevant tools to enhance patient education and support therapeutic interventions. These tools enable providers to translate medical concepts into experiential learning, thereby improving comprehension and treatment adherence.

VR headsets allow patients to enter immersive environments tailored to their care journey. A meta-analysis by Li et al. involving 878 participants found VR interventions improved comprehension by over 100%, reduced anxiety by 42%, and increased satisfaction [14]. Providers also use VR simulations to help patients visualize procedures or practice rehab exercises, making abstract instructions more concrete and actionable.

AR overlays digital elements on the real world. A scoping review of 11 AR-based medical education programs reported significant gains in knowledge and user satisfaction among learners [15]. In clinical settings, AR supports preoperative explanations, navigational overlays during consultations, and hands-on training scenarios that reinforce provider instruction.

Together, VR and AR enhance patient engagement by transforming complex health information into tangible experiences. These technologies alleviate cognitive overload, help patients prepare both physically and mentally, and support better adherence to treatment plans. Recent advances have lowered hardware costs and increased the availability of subscription-based VR and AR platforms, making these tools more feasible for clinics to adopt in patient education programs. As VR and AR technologies become more accessible, they integrate naturally with other advanced digital tools that enhance data analysis, personalization, and patient engagement. To fully unlock the potential of these immersive experiences, healthcare organizations require the high-speed, low-latency connectivity that 5G networks now make possible.

The Role of 5G in Healthcare Marketing

5G networks are reshaping healthcare marketing by enabling faster, more responsive patient engagement and strengthening telehealth delivery. With speeds up to 100 times faster than 4G and latency as low as 1 millisecond, 5G supports real-time interaction and improves digital communication between providers and patients [16, 17]. These performance figures apply under ideal conditions. For example, low latency supports patient-facing workflows such as high-resolution video education sessions or real-time remote rehabilitation check-ins. Implementing 5G in healthcare requires significant infrastructure investment, and security protocols must adapt to the higher volume and speed of transmitted data.

Real-Time Engagement and Immersive Experiences

5G enables marketers to deliver uninterrupted video consultations, real-time messaging, and immersive educational tools, such as augmented and virtual reality. These interactions create a more personalized and immediate experience for patients, which fosters trust and reinforces treatment adherence. For example, a clinic can utilize 5G-powered AR to guide a patient through a procedure during a consultation, enhancing comprehension and satisfaction.

Improved Telehealth and Mobile App Capabilities

5G improves the performance of mobile health apps and remote monitoring tools. Applications can now support advanced features like AI-based symptom checkers, real-time biometric tracking, and seamless EHR integration. This helps physicians monitor patients remotely with greater accuracy and fewer technical barriers. High-speed connections also reduce video lag, which is especially important when serving rural or underserved areas.

Strategic Marketing Implications

From a marketing perspective, 5G enables practices to deliver higher-quality content, such as streaming patient education videos without buffering. Faster response times and smoother user experiences can lead to higher completion rates for video visits and increased watch time for educational content, improving patient engagement metrics. Practices that adopt 5G-enabled tools gain a competitive advantage as patient expectations for convenience and digital access rise. Health systems can leverage these capabilities as differentiators in digital campaigns, positioning themselves as leaders in accessible, technology-driven care.

Big Data and Predictive Analytics

In healthcare, big data refers to complex and voluminous datasets that exceed the capacity of traditional data processing systems. These include structured and unstructured data from sources such as electronic health records (EHRs), medical imaging, wearables, genomics, clinical trials, insurance claims, and even patient-generated content from online platforms [18, 19].

This information provides a comprehensive overview of patient health trends, treatment patterns, and risk factors. By utilizing predictive analytics, healthcare systems can identify at-risk populations, anticipate surges in disease prevalence, and tailor interventions to individual or group characteristics. The accuracy of these predictions depends on data quality, and biased or incomplete datasets can lead to misleading insights and inappropriate outreach.

In one real-world example, Stanford University Medical Center implemented predictive modeling to identify heart failure patients at high risk for 30-day readmission.[7] After applying targeted post-discharge interventions, readmission rates dropped from 21% to 15%, representing a 29% relative reduction. This illustrates how data-informed outreach can improve patient outcomes and reduce avoidable returns.

Targeting Patients with Personalized Campaigns

Healthcare marketers now utilize predictive analytics to segment audiences by risk profile, demographic traits, and behavioral patterns. For instance, if data indicates a rising incidence of prediabetes in a specific zip code, providers can launch targeted educational campaigns focused on nutrition, exercise, and glucose screening in that area. These targeted efforts enhance both engagement and outcomes.

Predictive modeling also enables marketers to time their outreach more effectively. Instead of waiting for patients to seek help, organizations can deliver messages that anticipate needs, such as flu vaccine reminders ahead of seasonal outbreaks or appointment nudges for high-risk patients overdue for screenings.

Planning and Forecasting

The ability to forecast health events transforms marketing into a proactive strategy. Campaigns can be aligned with predicted trends, resource needs, or service utilization spikes. This allows providers to manage demand more efficiently, reduce no-show rates, and strengthen patient relationships.

Ethical and Privacy Considerations

While the potential of big data is significant, it raises ethical concerns. Marketers must ensure that data use complies with HIPAA and other data protection regulations. Patients should be informed when their data is used for segmentation or targeting. Explicit consent, clear privacy policies, and algorithmic fairness are essential to maintaining trust.

Predictive analytics must also avoid reinforcing bias. Algorithms trained on historical data may inadvertently replicate existing disparities. Oversight is necessary to ensure equitable messaging and prevent excluding vulnerable groups from outreach efforts.

---

[7] https://revealhealthtech.com/blogs/beyond-discharge-how-predictive-models-are-transforming-readmission-rates/?utm_source=chatgpt.com. Accessed 8/9/25.

Taken together, these technologies form the foundation for the next wave of healthcare marketing innovations, which are beginning to merge and reinforce each other in practice.

Sidebar 2: What Makes Patient Data Use HIPAA-Compliant in Marketing?

When using patient data in healthcare marketing, HIPAA sets strict rules to protect privacy. Follow these practices to remain compliant:

Consent

- Obtain explicit patient consent before using any identifiable information for marketing outside of treatment, payment, or healthcare operations.
- Avoid using protected health information (PHI) for targeted advertising unless this use is disclosed in the Notice of Privacy Practices and acknowledged by the patient.

Data Handling

- De-identify data according to HIPAA standards:
- **Safe Harbor**: Remove all 18 identifiers (e.g., name, email, birth date, zip code).
- **Expert Determination**: Have a qualified statistician certify minimal reidentification risk.
- Avoid technologies that may collect PHI without consent, such as tracking pixels embedded in patient portals.

Tool Compliance

- Use HIPAA-compliant platforms for customer relationship management (CRM), analytics, and email marketing.
- Ensure all vendors handling PHI have signed Business Associate Agreements (BAAs).

Documentation and Transparency

- Keep records of how patient data is collected, processed, and secured.
- Be transparent about data practices in all patient-facing communications.
- Review state privacy laws, as they may impose stricter requirements than HIPAA.

Following these steps reduces regulatory risk and helps maintain patient trust.

Future Outlook and Emerging Innovations

Digital healthcare marketing continues to evolve, shaped by advancements in artificial intelligence (AI), big data analytics, wearable technologies, and remote monitoring systems. These tools are no longer theoretical; they are embedded in patient communication strategies and reshaping marketing frameworks in clinical settings. Figure 10.1 summarizes these core emerging technologies and illustrates how each contributes to patient engagement, personalization, and campaign effectiveness.

1. **Artificial Intelligence (AI) in Marketing and Patient Targeting**
   AI tools are increasingly used to identify patient preferences, automate message personalization, and segment populations based on clinical or behavioral

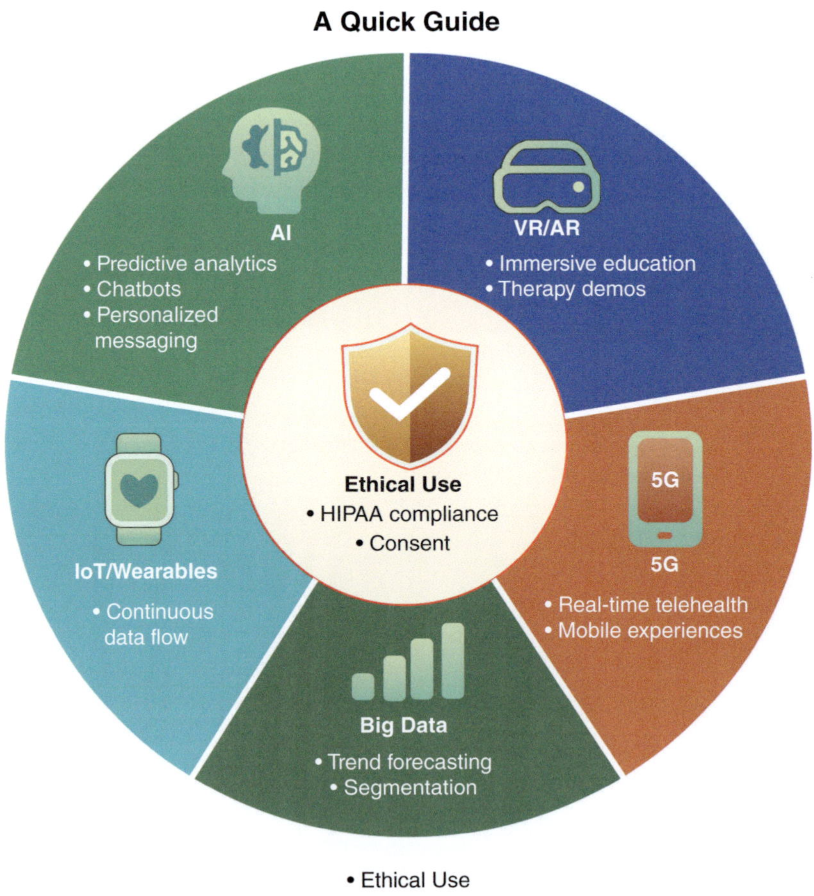

Fig. 10.1 This visual summarizes key emerging technologies shaping digital healthcare marketing, including AI, VR/AR, 5G, predictive analytics, and wearable tech. It illustrates how each tool enhances patient engagement, personalization, and campaign effectiveness. (Image by Quinn/ChatGPT)

risk profiles. For example, health systems are integrating AI into customer relationship management platforms to customize outreach and reduce redundant touchpoints. Research shows that organizations with structured AI governance and cross-functional coordination are more likely to translate AI capabilities into actionable business strategies [20].

2. **Telehealth and Remote Monitoring**
   The expansion of telehealth during the COVID-19 pandemic normalized virtual care and redefined the role of marketing. Healthcare organizations now use digital campaigns to guide patients through telemedicine services and remote care options. These platforms also generate engagement data that feeds back into marketing algorithms. As broadband and mobile connectivity improve, these channels are expected to dominate outpatient engagement [21].
3. **Wearable Devices and Predictive Engagement**
   Wearable health technologies and IoT-connected monitors contribute new streams of data for marketers and care teams. These data, when analyzed, support predictive messaging, such as reminders for preventive screenings or adherence nudges. This integration enables more proactive outreach and aligns marketing messages with real-time health behaviors [22].
4. **Ethics, Consent, and Algorithmic Transparency**
   As marketing efforts rely more on AI-driven tools, ethical use of patient data remains a top priority. Consent must be clearly obtained, and models should be audited for bias. Without transparency and explainability, AI-based outreach can erode patient trust and lead to regulatory challenges. Recent research emphasizes the importance of fairness and accountability in health AI deployments, particularly those involving marketing and communication tools [23, 24].
5. **Organizational Agility and Metrics-Driven Impact**
   Healthcare organizations that embed AI into marketing functions benefit most when they operate with cross-functional teams and continuous learning structures. Leaders must define measurable success criteria early—such as appointment adherence, new patient acquisition, or engagement rates—to evaluate the impact of digital outreach strategies [12]. This ensures that innovation efforts produce demonstrable clinical and business value.

Sidebar 3: How to Vet Emerging Tech Vendors

Before adopting a new digital health or marketing technology, conduct due diligence to ensure safety, compliance, and clinical relevance:

- **Request outcome data**: Ask for evidence of real-world performance. Insist on metrics tied to clinical or operational improvements, not just feature lists or demo videos.
- **Security:** Confirm the vendor's security posture by reviewing SOC 2 Type II certification, recent penetration test results, and uptime guarantees to ensure data protection and service reliability.

(continued)

- **Verify HIPAA compliance**: Ask for documentation of HIPAA safeguards, including encryption practices and Business Associate Agreements (BAAs).
- **Clarify data ownership**: Determine who controls and stores patient data. Confirm whether the vendor shares or sells de-identified data to third parties.
- **Check for EHR interoperability**: Ensure the platform can integrate with your EHR system or practice management tools using standard protocols like HL7 or FHIR.

A vendor with flashy AI claims but no track record, regulatory footing, or integration path will likely waste your time and possibly compromise patient trust.

Conclusion

Emerging technologies are reshaping healthcare marketing. Artificial intelligence, predictive analytics, augmented and virtual reality, SMS platforms, and 5G networks are no longer experimental. They now influence how physicians reach patients, deliver education, and improve engagement. These tools allow you to target communication more precisely, automate outreach, and adapt content in real time based on patient behavior or risk.

Effective digital marketing now depends on your ability to integrate these technologies while maintaining ethical standards, data transparency, and measurable outcomes. As the landscape evolves, healthcare providers who stay informed and adaptive will be better positioned to attract, engage, and retain patients in a competitive digital environment.

Next Steps

1. Select one patient messaging workflow to automate with AI, and measure improvement in appointment confirmation rates after 90 days.
2. Choose one analytics use case, such as campaign click-through tracking, and set a goal for a measurable increase in conversion within 90 days.
3. Pilot one VR or AR initiative for patient education, with the aim of achieving a set percentage increase in post-session knowledge scores after 90 days.

References

1. Clerk Chat. Uses of AI in healthcare marketing for patient engagement and ROI. Clerk Chat. 2024. https://clerk.chat/blog/ai-in-healthcare-marketing. Accessed 1 Aug 2025.
2. ZealousWeb. Reduce no-shows with WhatsApp & SMS reminders. ZealousWeb. 2025. https://www.zealousweb.com/blog/reduce-no-shows-appointment-reminders-whatsapp-sms. Accessed 1 Aug 2025.
3. Lakhani, Nina Text reminders could save NHS millions – Health News – Health & Families. The Independent. 2011. Accessed 9 Aug 2025.
4. McLean SM, Booth A, Gee M, Salway S, Cobb M, Bhanbhro S, Nancarrow SA. Appointment reminder systems are effective but not optimal: results of a systematic review and evidence synthesis employing realist principles. Patient Prefer Adherence. 2016;10:479–99. https://doi.org/10.2147/PPA.S93046. PMID: 27110102; PMCID: PMC4831598.
5. Sims H, Sanghara H, Hayes D, Wandiembe S, Finch M, Jakobsen H, Tsakanikos E, Okocha CI, Kravariti E. Text message reminders of appointments: a pilot intervention at four community mental health clinics in london. Psychiatr Serv. 2012;63(2):161–8.
6. Graphlogic AI. Conversational AI for healthcare: from missed appointments to meaningful engagement. Graphlogic AI. 2025. https://graphlogic.ai/blog/case/conversational-ai-for-healthcare. Accessed 1 Aug 2025.
7. Innovation VA. Immersive virtual reality in health care literature compendium [PDF report]. U.S. Department of Veterans Affairs. 2024. https://www.innovation.va.gov. Accessed 1 Aug 2025.
8. Triton Ong, M., Ruppert, M., Rashidi, P., Ozrazgat-Baslanti, T., Bihorac, A., & Suvajdzic, M. (2019). The DREAMS project: improving the intensive care patient experience with virtual reality. arXiv. https://arxiv.org/abs/1906.11706. Accessed 1 Aug 2025.
9. Mergen M, Graf N, Meyerheim M. Reviewing the current state of virtual reality integration in medical education – a scoping review. BMC Med Educ. 2024;24:788. https://doi.org/10.1186/s12909-024-05777-5.
10. Elendu C, Amaechi DC, Okatta AU, Amaechi EC, Elendu TC, Ezeh CP, Elendu ID. The impact of simulation-based training in medical education: a review. Medicine (Baltimore). 2024;103(27):e38813. https://doi.org/10.1097/MD.0000000000038813. PMID: 38968472; PMCID: PMC11224887.
11. Naeem M, Alkhadrawi R, Ashraf H, Ibrahim MT. Exploring the potential of artificial intelligence and machine learning to combat COVID-19 and existing opportunities for LMICs: a scoping review. Primary Care Commun Health. 2020;11:1–8. https://doi.org/10.1177/2150132720963644.
12. Buch VH, Ahmed I, Maruthappu M. Artificial intelligence in healthcare: past, present and future. Future Healthcare J. 2018;5(2):94–8. https://doi.org/10.7861/futurehosp.5-2-94.
13. Jha S, Topol EJ, Williams D. Artificial intelligence in healthcare: transforming the practice of medicine. Future Healthcare J. 2021;8(2):e188–93. https://doi.org/10.7861/fhj.2021-0095.
14. Li Y, Kim M, Sutedjo A. Effects of virtual reality as a patient education tool: a meta-analysis. J Educ Technol Develop Exchange. 2024;17(2):216–28. https://doi.org/10.18785/jetde.1702.10. Accessed 1 Aug 2025
15. Barsom EZ, Graafland M, Schijven MP. Systematic review on the effectiveness of augmented reality applications in medical training. BMJ Open. 2016;10(9):e038416.
16. George LE, Georgekutty R, Sahaya BK. 5G use in healthcare: the future is present. NIH Res Rev Health Sci. 2021;25(4):2021.00006. https://doi.org/10.21601/NRR.2021.00006. PMID: 34807636; PMCID: PMC8765483.
17. Dewji S, Deshmukh K, Joseph N, Alzahrani M, Ali M, Sondhi Y. 5G Technology in Healthcare and Wearable Devices: a review. *Biomedical engineering*. International. 2021;2021(6):142–50. PMID: 35074471; PMCID: PMC9007585
18. Batko K, Ślęzak A. The use of big data analytics in healthcare. Big Data. 2022;10(2):97–106. https://doi.org/10.18686/bd.v10i2.60543. PMID: 35101479; PMCID: PMC8794712

19. Ristevski B, Chen M. Big data analytics in medicine and healthcare. J Integr Bioinform. 2018;15(3):20170030. https://doi.org/10.1515/jib-2017-0030. s
20. Kim H, Chen Y, Shah RN, Zhou A, Kelly CJ, Shah NH. Organizing AI in healthcare: cross-functional teams, system integration, and data governance. arXiv preprint arXiv. 2023:2211.06346. https://doi.org/10.48550/arXiv.2211.06346.
21. Budd J, Miller BS, Manning EM, Lampos V, Zhuang M, Edelstein M, et al. Digital technologies in the public-health response to COVID-19. Nat Med. 2020;26(8):1183–92. https://doi.org/10.1038/s41591-020-1011-4.
22. Reddy S, Rogers W, Wiles L. Predictive analytics in health care: an overview of current trends and challenges. J Med Syst. 2022;46(4):1–10. https://doi.org/10.1007/s10916-022-01748-6.
23. Pasricha AK, Aggarwal A, Joshi R, Mukhopadhyay S. AI ethics in smart healthcare: fairness, accountability, and transparency. arXiv preprint arXiv. 2022:2208.05437. Accessed 1 Aug 2025
24. Reuters. Health rounds: AI can have medical care biases too, study reveals. Reuters Health. 2025. https://www.reuters.com/business/healthcare-pharmaceuticals/health-rounds-ai-can-have-medical-care-biases-too-study-reveals-2025-04-09/. Accessed 1 Aug 2025.

Chapter 11
Ethical Considerations and Patient Privacy in Digital Marketing for Healthcare

Introduction to Ethical Considerations in Digital Marketing

Digital marketing gives healthcare providers new ways to reach patients, including email, social media, search targeting, and AI. But these strategies also raise ethical concerns about truthfulness, consent, and data use.

Marketers must avoid overstating benefits or downplaying potential harms. For instance, nearly nine out of ten beauty clinics in the United Kingdom violated advertising laws by exaggerating the effectiveness of Botox and other cosmetic procedures, often ignoring or concealing potential side effects such as pain, bruising, or contraindications. These ads were so misleading that regulatory authorities initiated bans or enforcement actions [1].

Safeguarding patient data remains equally important. As marketing teams incorporate digital platforms and analytics, they must rely only on de-identified or fully consented protected health information (PHI). Tools must meet HIPAA standards, and patients should know how their data will be used and have the option to opt out.

In 2022, an investigation by *The Markup* and *STAT News* revealed that hundreds of US hospitals were using the Meta Pixel tracker on their websites, including on pages where patients scheduled appointments or looked up sensitive conditions [2]. This tool, often used to optimize digital marketing, sent protected health information (PHI) such as IP addresses, appointment types, and search queries directly to Facebook, without patient knowledge or consent.

This breach not only violated HIPAA but also triggered class-action lawsuits and investigations by regulators. Some hospitals were forced to disable the pixel and reevaluate their entire digital marketing infrastructure to avoid further legal exposure and loss of patient trust.

Ethical marketing in healthcare demands honesty, patient autonomy, and data protection. Providers must stay aligned with changing legal and professional norms.

C. Quinn, *The Digital Practice*, https://doi.org/10.1007/978-3-032-07745-5_11

To understand these responsibilities, it's helpful to examine how digital platforms have changed patient privacy expectations.

Understanding Patient Privacy

Patient privacy was once guided primarily by professional discretion, focused on physical records and face-to-face encounters. Today, digital platforms have redefined those boundaries. Privacy now encompasses data shared across electronic health records (EHRs), patient portals, telehealth sessions, and even social media interactions. Patients expect control over their personal health information (PHI) and personally identifiable information (PII) shared with healthcare organizations, insurers, and third-party vendors.

Healthcare marketers must treat patient data with the same caution as clinicians. Every click, form submission, or online message could involve PHI.

In a 2022 study published in *Health Affairs*, researchers found that 99% of US hospital websites transmitted visitor data to third parties through trackers like Meta Pixel or Google Analytics, sometimes including information about search terms, scheduled appointments, or logged-in patient activity [3]. In response, several hospitals faced lawsuits and regulatory scrutiny over unauthorized data sharing with advertisers [4].

These incidents underscore a larger problem. Privacy expectations have risen, but enforcement often lags. A Pew Research Center survey found that 81% of Americans feel they have little control over the data companies collect about them [5].

As the boundaries between health and tech blur, marketers must avoid vague promises and casual data use. Trust, once broken, is difficult to repair. These shifting expectations highlight why strict legal compliance, particularly under HIPAA, is not just a formality but a safeguard for patient trust.

HIPAA and Digital Marketing Compliance

The Health Insurance Portability and Accountability Act (HIPAA) sets national standards for protecting patient health information. Enacted in 1996 and strengthened by the HITECH Act, HIPAA requires covered entities and business associates to safeguard protected health information (PHI) against unauthorized access or disclosure [6].

Healthcare marketers must avoid digital tools that collect or transmit PHI without proper controls. Embedding tools like Meta Pixel or Google Analytics on patient-facing pages may inadvertently share sensitive data with third parties; this behavior has triggered investigations, class-action lawsuits, and scrutiny from regulators [7].

Key HIPAA Requirements for Marketing Use

- Written patient authorization is mandatory if PHI is used for any marketing outreach outside routine healthcare operations.
- Use only HIPAA-compliant platforms. Ensure all third-party vendors sign a Business Associate Agreement (BAA). These agreements extend HIPAA responsibilities to partners that handle ePHI [8, 9]. For a detailed list of HIPAA-compliant marketing tools, including secure email platforms, encrypted texting services, analytics tools that protect PHI, and CRM systems designed for healthcare, see Appendix 11.4: Recommended HIPAA-Compliant Tools for Marketing.
- De-identified data must meet either Safe Harbor or Expert Determination standards to avoid reidentification risk.
- Implement technical safeguards, including encryption, access controls, and audit logs, consistent with the Security Rule.

Violations carry serious penalties. Civil fines range from $141 to $2.13 million per violation, depending on culpability, and may result in formal corrective action plans. Business associates can be penalized separately under HIPAA/HITECH enforcement [10, 11]. For a quick-reference guide explaining what qualifies as PHI, when patient consent is required, and when Business Associate Agreements are necessary, see Appendix 11.1 HIPAA and Marketing: Quick Guide.

To apply HIPAA in real-world marketing contexts, teams must know which types of data count as PHI and what that means in practice. Additionally, while HIPAA provides the legal foundation, informed consent ensures patients remain active participants in decisions about their data.

Sidebar: What Counts as PHI in Marketing?
When healthcare marketing involves any of the following data linked to an individual, it may qualify as protected health information (PHI) under HIPAA:

- **Names** (first or last name with medical context)
- **Email addresses** tied to appointment requests or health inquiries
- **Phone numbers** used in patient communications
- **IP addresses** collected from web forms or portals
- **Dates** (e.g., appointment dates, birthdates, discharge dates)
- **Device identifiers** from patient portals or mobile health apps
- **Geolocation data** that can identify a patient's residence or visit history
- **Health conditions, diagnoses, or treatment details** used in targeting
- **Photos or videos** of patients in any marketing material
- **Any data combined with health service details** (e.g., "John signed up for a diabetes seminar")
- **Key Rule**

(continued)

If the data can be used to identify a person *and* it relates to health services, payment, or medical history, it is PHI, even if it's collected through a website form, chatbot, or email campaign.
Use only HIPAA-compliant tools when handling this data, and never use it for marketing without explicit written consent.

Informed Consent in Digital Communications

Informed consent in healthcare marketing means clearly explaining to patients how their data will be used and obtaining their agreement before proceeding. This goes beyond a checkbox. It involves giving patients information they can understand so they can make informed choices about participation in digital tools, health apps, or marketing programs. Effective consent not only shows respect for patient autonomy but also protects providers and marketers from regulatory risks.

A 2020 study highlighted gaps in comprehension when patients were asked to agree to data-sharing policies embedded in apps or websites. Many users skimmed or skipped privacy statements, and few retained critical details, such as whether their data would be shared with third parties [12]. In one case, a health app collected PHI and shared it with advertisers. This is despite having a consent checkbox that patients didn't check because the language used was vague and hard to understand [13].

To reduce these risks and respect patient rights, health marketers should adopt robust consent practices. Once consent protocols are in place, marketers must apply them consistently across every aspect of campaign development, from messaging to media use. For a ready-to-use sample of plain-language consent text that meets HIPAA requirements and is designed for websites, patient portals, and other digital platforms, see Appendix 11.2: Sample Consent Language for Digital Platforms.

See Fig. 11.1 for an example of how informed consent can be documented on digital platforms in a way that meets HIPAA requirements and is easy for patients to understand.

Fig. 11.1 Documenting informed consent on digital platforms (Image by Quinn/ChatGPT)

Ethical Marketing Strategies

Ethical healthcare marketing begins with protecting patient privacy and continues with clear and honest communication. This section outlines practical steps that help organizations avoid legal risks and maintain public trust.

Respect for Patient Privacy

Marketers must obtain written permission before using any patient content, including photos, quotes, or videos. A consent form should explain how the material will appear, when it will be used, and how long the permission lasts. De-identifying patient stories does not excuse the use of recognizable images or quotes. Avoid pressure tactics. For example, stay away from phrases like "limited time only" or "don't miss out." Give patients facts. Let them decide without emotional manipulation.

Clear and Honest Messaging

Patients deserve accurate information. Avoid vague claims such as "best in class" or "top-rated care." Explain treatment risks. Share evidence when promoting outcomes. Avoid omitting important facts or exaggerating benefits. If your practice offers a new therapy, explain how it compares with standard care. Build trust by being specific, truthful, and easy to understand.

Callout Box: Signs of Unethical Marketing

- Claims a guaranteed outcome.
- Shows patient photos or quotes without signed consent.
- Hides treatment risks or side effects.
- Removes negative reviews.
- Fails to mention the provider's business ties.
- Says "#1 rated" without proof.
- Markets unapproved treatments without stating that status.
- Uses guilt, fear, or urgency to push decisions.

Social media presents unique ethical risks because content often spreads rapidly and outside formal approval workflows. For a printable checklist of common ethical red flags in healthcare marketing, with explanations of why each is problematic, see Appendix 11.3: Ethical Marketing Red Flags Checklist.

Social Media Ethics

Healthcare marketers rely on social media to reach patients, but these platforms bring ethical challenges. When promoting care or services, teams must protect patient privacy and avoid misleading or exploitative content.

Engagement Without Compromise

Healthcare organizations can share patient stories, but only after removing identifiable details and securing written consent. These safeguards protect patient privacy while still allowing storytelling.

For example, a children's hospital campaign on Instagram featured parents' testimonials about pediatric cancer care. The hospital obtained consent, anonymized all details, and posted only after final approval from the families. This approach built trust and avoided privacy violations. The campaign won a PRNEWS award for its storytelling approach while protecting privacy [14].

Even with such precautions, marketers must remain vigilant to avoid unintentional disclosures that breach patient trust.

Confidentiality Comes First

HIPAA rules still apply on social platforms. Posting clinical images, appointment details, or even vague references to a case may violate confidentiality if someone can identify the patient. Marketing teams must review all content and confirm no links to patient identity exist unless explicit permission was granted.

Ethics in Everyday Posts

Healthcare marketers should never exaggerate results, omit risks, or post clickbait. If a campaign highlights a new treatment, it must mention risks or limitations. If a video shares a recovery story, it must state that the individual's experience may not apply to others. Transparency prevents harm and helps patients make informed choices.

Avoid claims like "guaranteed cure" or "best in city" unless supported by published data. Always state potential risks or limitations of any treatment or procedure. Patients deserve clear information, not marketing hype.

Skip countdown timers, guilt-based messaging ("don't delay your treatment"), or emotional manipulation. Those tactics undermine patient autonomy and compromise ethical standards.

Use Social Media Responsibly

Effective social media campaigns inform the public, support health literacy, and strengthen community ties. Ethical marketers publish accurate content, safeguard privacy, and secure informed consent before sharing stories. These practices protect patients and uphold the credibility of healthcare institutions.

While messaging and media are critical, none of it matters without a secure infrastructure to protect the patient data behind those campaigns.

See Fig. 11.2 for a summary of the types of content healthcare professionals should avoid posting on social media to protect privacy and maintain trust.

Digital Marketing and Patient Data Security

Securing patient data is essential in digital healthcare marketing. This section outlines key challenges and practical strategies to protect sensitive information.

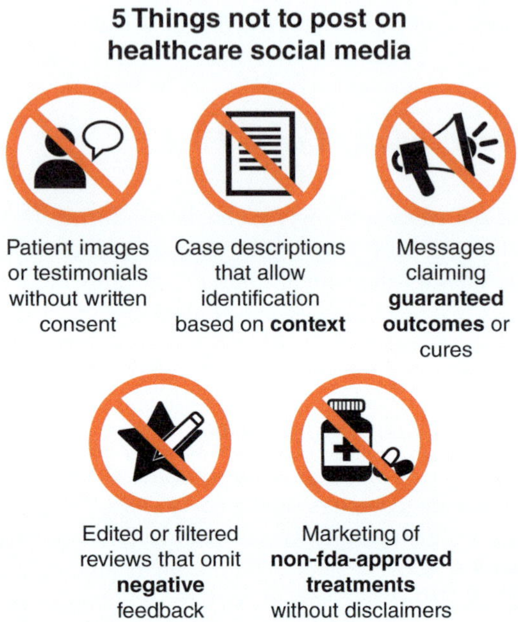

Fig. 11.2 Healthcare professionals should avoid posting five types of content on social media to protect patient privacy, comply with regulations, and maintain public trust. (Image by Quinn/ChatGPT)

Challenges in Securing Patient Data

Healthcare digitalization has increased the risk of unauthorized access to personal health information. Cyber threats are becoming more sophisticated, and the platforms that store and transmit patient data are often complex and difficult to secure. These vulnerabilities raise concerns about both privacy and regulatory compliance.

Best Practices for Data Security

To safeguard data, healthcare organizations should adopt a structured approach to security. The following practices help reduce risk and promote trust.

Comprehensive Security Policies

Develop clear security policies that regulate access to patient data. These policies must assign role-based access using the principle of least privilege. A nurse might access clinical records but not billing data. An administrator might see financial details but not health records. These distinctions help prevent unauthorized use. Organizations must review and update these policies frequently to respond to evolving threats. Following standards like the HIPAA Security Rule helps ensure compliance.

Data Encryption

Encrypt all patient data in storage and in transit. Encryption ensures that if data is intercepted or stolen, it remains unreadable without the correct decryption key.

Data at Rest

"Data at rest" refers to information that is stored and not actively moving through a network. In healthcare, this includes patient records saved in databases, on hard drives, or in cloud storage systems. To protect this stored information, organizations should use encryption methods such as AES-256.

AES stands for Advanced Encryption Standard, a widely recognized method for securing data. The "256" indicates the strength of the encryption key, which has 256 bits, making it extremely difficult to break. This standard is endorsed by the National Institute of Standards and Technology (NIST) and is used by government agencies and healthcare institutions to safeguard sensitive information.

Encrypting data at rest ensures that even if someone gains unauthorized access to the storage system, they cannot view the contents without the proper decryption key. For example, a hospital that uses AES-256 encryption on its internal servers can keep patient records unreadable to hackers or unauthorized staff.

Data in Transit

"Data in transit" refers to patient information that is actively moving between systems. This includes data transferred within hospital networks, between a clinician's device and an electronic health record (EHR) system, or through Internet-based communication. Because information in motion is especially vulnerable to interception, strong encryption is essential to protect its confidentiality.

Healthcare organizations use Transport Layer Security (TLS) to safeguard data during transmission. TLS is a widely accepted encryption protocol that ensures the data is unreadable while it moves across networks. When a clinician accesses patient records remotely, TLS encrypts the information before it leaves the hospital server and decrypts it only after it reaches the intended device. This process begins with a secure handshake, where the sender and recipient exchange cryptographic keys. These keys are used to lock and unlock the data. Even if a hacker intercepts the transmission, the encrypted information remains unreadable without the decryption key, which is only available to the authorized recipient.

Organizations such as the National Institute of Standards and Technology (NIST) and the Health Information Trust Alliance (HITRUST) recommend the use of TLS as part of a broader strategy to ensure data security and regulatory compliance [15, 16].

Regular Security Audits

Conduct routine audits to detect technical and human vulnerabilities. A technical audit might scan the EHR system for outdated software, weak passwords, or unpatched vulnerabilities. A human factor audit might review user permissions, simulate phishing attacks, and assess staff training. Based on audit results, the organization should strengthen access controls, enhance staff education, or update encryption methods. These steps align with HITRUST CSF and NIST SP 800–53 frameworks [15, 16].

Training and Awareness

Training healthcare staff on cybersecurity is one of the most effective ways to reduce data breaches caused by human error. Regular, targeted education helps employees recognize threats, understand their responsibilities, and respond appropriately to potential incidents.

Staff should be trained to identify phishing emails, avoid suspicious links, secure mobile devices, use strong passwords, and report security incidents immediately. Training programs should also include simulation exercises such as phishing tests, which reinforce learning and help administrators track improvement over time.

Frequent refreshers improve retention and foster a culture of vigilance. According to the *Healthcare Information and Management Systems Society* (HIMSS), 67% of healthcare cybersecurity incidents in recent years involved phishing attacks, which are often successful due to insufficient staff training [17].

A 2022 report by the *Ponemon Institute* found that hospitals with formal cybersecurity training programs experienced 43% fewer data breaches caused by human error compared to those without such programs [18].

Effective programs should align with frameworks such as the NIST Cybersecurity Framework or HITRUST CSF, which recommend role-based training tailored to specific job functions.

By investing in continuous, evidence-based training, healthcare organizations can significantly reduce their exposure to preventable cybersecurity threats and ensure that every employee plays a role in protecting patient data. By implementing these strategies, healthcare organizations can improve data security, meet compliance standards, and maintain patient trust.

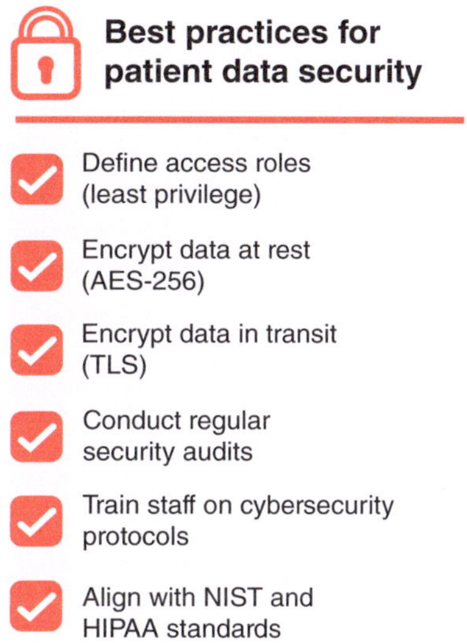

Fig. 11.3 Checklist of best practices for patient data security, including access control, encryption, regular audits, staff training, and alignment with regulatory standards. (Image by Quinn/ChatGPT)

Even with strong technical defenses, marketing teams must still navigate complex ethical dilemmas that arise when legal boundaries are unclear.

See Fig. 11.3 for a checklist of best practices that strengthen patient data security and support compliance with HIPAA and other regulations.

Ethical Gray Areas in Digital Marketing

Healthcare marketers often face difficult ethical choices, especially when digital strategies intersect with patient privacy. One recurring gray area involves the use of patient data for targeted advertising. While this data can improve marketing precision, it also raises concerns about consent, data protection, and the possibility of exploiting vulnerable individuals.

Real-World Dilemma: Geotargeting without Consent

In 2022, a mental health clinic launched a digital ad campaign using geotargeting to reach individuals near psychiatric hospitals. This case involved a marketing firm that used smartphone geolocation data to deliver addiction treatment ads to individuals who had visited methadone clinics and other addiction recovery centers.

Although technically legal, the campaign drew criticism for targeting users based on inferred health status without explicit consent [19].

The backlash was swift: patients complained, advocacy groups intervened, and the clinic faced reputational damage. The Massachusetts Attorney General investigated the case, prompting calls for tighter regulation. After internal review, the clinic suspended the campaign and instituted a new policy requiring ethics committee approval for all ads involving patient data.

They also adopted stricter data use protocols and improved consent mechanisms. The lesson: even when the law permits a practice, ethical standards must guide final decisions.

With these principles in mind, organizations can design marketing approaches that strengthen engagement while protecting privacy.

When Laws Lag Behind Technology

Technology often moves faster than regulation. In situations where legal guidance is unclear, healthcare organizations should follow the precautionary principle: when in doubt, err on the side of patient autonomy, data minimization, and informed consent. Until laws catch up, organizations can consult voluntary frameworks like the AMA Code of Medical Ethics or the FTC's guidance on data use. Internal ethics committees, legal counsel, and cross-functional review teams can help bridge the gap between innovation and ethical restraint.

By proactively addressing these gray areas, organizations can protect patients, avoid public backlash, and preserve institutional trust. Ethical marketing is not about avoiding risk. It is about earning credibility by making patient-centered decisions—even when the law stays silent. One way to address these uncertainties is by centering patient values in all digital communication strategies.

Patient-Centric Marketing: Balancing Engagement and Privacy

Creating healthcare marketing strategies that engage patients while respecting their privacy is both a challenge and an opportunity for healthcare marketers. A well-designed, patient-centric marketing campaign recognizes the importance of transparency, consent, and data security, ensuring that all marketing efforts comply with legal regulations and prioritize patient confidentiality. Meeting these expectations requires a firm grasp of both existing legal standards and the emerging regulations shaping healthcare marketing.

Building Privacy-Respectful, Patient-Centered Strategies

1. **Listen First: Understanding Patient Needs**

   Effective marketing begins by understanding the patient's journey, challenges, and needs. Engaging patients through surveys, feedback forms, and social listening tools helps healthcare organizations tailor campaigns to resonate with real patient concerns. This creates a relationship built on empathy and trust.

   Studies show that listening to patient feedback not only strengthens engagement but can reduce dissatisfaction—80% of patients say that when healthcare providers listen to their concerns, they feel more valued and satisfied with the care provided.[1,2]

2. **Tailored Campaigns with Clear Boundaries**

   Personalized campaigns improve both engagement and privacy. For instance, a hospital specializing in diabetes care could create specific educational content for newly diagnosed patients. This content could include treatment options, lifestyle management tips, and invitations to support groups. By focusing on a specific audience, marketers can ensure that only the necessary data is used while still providing useful information to patients. This targeted approach aligns with ethical principles by minimizing the over-collection of sensitive data. Privacy concerns are significantly reduced, and patients feel more informed about their healthcare choices.

3. **Combining Data Analytics with Empathy**

   Data analytics enables healthcare marketers to understand patient behavior and preferences, but these insights should always be paired with empathy. For example, a physical therapy clinic could use patient recovery data to design personalized email campaigns for knee surgery patients. Early-stage patients might receive pain management tips, while later-stage patients could receive strengthening exercises. These communications are crafted with empathy, acknowledging the patient's recovery challenges and offering practical support. To ensure privacy, patient data must be anonymized and stored securely, complying with HIPAA regulations.

[1] https://www.regal.ai/patient-experience-glossary-terms/how-does-active-listening-result-in-better-patient-care-glossary. Accessed 8/9/25.

[2] https://www.techtarget.com/patientengagement/news/366584982/What-Are-Patient-Perceptions-Experiences-with-Value-Based-Care. Accessed 8/9/25.

Techniques for Personalizing Marketing Efforts Without Compromising Patient Confidentiality

- **Anonymized Data**
  Anonymized patient data offers valuable insights into patient preferences without compromising privacy. Aggregating data in this way allows marketers to segment populations and tailor content without identifying individuals. This approach aligns with privacy regulations while enhancing marketing effectiveness.
- **Secure Communication Channels**
  Marketing communications should always be delivered through encrypted, HIPAA-compliant channels. These tools safeguard sensitive information and ensure that communications meet healthcare standards for patient privacy. Utilizing platforms like [insert HIPAA-compliant tools] ensures that data is not exposed to unauthorized parties.
- **Require Explicit Consent**
  Consent is the cornerstone of patient privacy. Marketers must obtain explicit consent from patients before using their data for marketing purposes. This consent should be informed, voluntary, and revocable. Transparency about data usage builds trust, and allowing patients to opt out at any time is crucial.

Real-World Example: A Healthcare System Navigating the Privacy-Engagement Balance

Case: *Memorial Sloan Kettering Cancer Center (MSKCC)—Precision Engagement with Privacy Controls*

MSKCC has implemented patient engagement campaigns using secure, HIPAA-compliant platforms such as Salesforce Health Cloud and encrypted email to deliver personalized educational content to oncology patients.

By utilizing data analytics, the marketing team identified patients who had undergone recent screenings but had not yet attended follow-up care. These patients had to opt in to receive communications.

The hospital sent tailored educational content about treatment options and available support services. Data used in the campaign was anonymized and securely stored. Consent was obtained from patients before any communication, and feedback was gathered to ensure the campaign's relevance and effectiveness. This approach helped boost patient follow-up appointments by 40% while maintaining privacy and trust [20].

Balancing patient engagement with privacy is a delicate task, but when done correctly, it fosters trust and enhances the effectiveness of healthcare marketing. By listening to patients, respecting their privacy, and using data responsibly, healthcare organizations can create marketing campaigns that not only drive engagement but also build long-lasting relationships based on transparency and respect.

Patient-centered marketing works best when it operates within a clear legal framework, which continues to evolve as new risks emerge.

Legal and Ethical Compliance in Digital Marketing

Healthcare digital marketing requires strict adherence to legal standards designed to protect patient privacy, ensure content accuracy, and maintain ethical integrity. The most prominent regulatory frameworks are the Health Insurance Portability and Accountability Act (HIPAA) and the Federal Trade Commission (FTC) Act.

HIPAA and Healthcare Marketing

The HIPAA Privacy Rule prohibits the unauthorized use of protected health information (PHI) in marketing communications. Healthcare marketers must obtain prior written authorization from patients before using their data for promotional campaigns. This includes emails, newsletters, or advertising that references specific health conditions or treatments [21]. Any failure to safeguard PHI or use it without consent can lead to civil penalties, reputational damage, and legal liability. In 2023, the Office for Civil Rights fined a medical practice $300,000 for using patient testimonials in marketing without proper authorization [22].

FTC Oversight and Advertising Accuracy

The FTC Act prohibits unfair or deceptive advertising, including false claims about medical treatments or technologies. All health-related marketing materials must be truthful, not misleading, and supported by scientific evidence [23]. For example, in 2022, the FTC sanctioned a telehealth company for making unsubstantiated claims about the effectiveness of its anxiety treatment program. The agency required the company to retract the claims and implement compliance monitoring procedures.

Emerging Risks: AI and Privacy Regulation

As healthcare marketers adopt AI tools to automate campaigns and personalize messaging, they face new risks. AI-generated content must be audited for accuracy and bias. When using AI-driven targeting, marketers must ensure transparency about how patient data is used and avoid drawing inferences that violate privacy laws. HIPAA and the FTC have both issued warnings about the misuse of machine learning tools in healthcare advertising [24].

Additionally, healthcare marketers should monitor state-level legislation like the California Consumer Privacy Act (CCPA), which gives patients the right to opt out of data sharing and requires disclosure of data use practices. For international marketing campaigns, compliance with the EU's General Data Protection Regulation (GDPR) is also critical.

By aligning digital strategies with legal requirements and ethical expectations, healthcare organizations can avoid regulatory scrutiny and strengthen public trust.

Compliance Checklist for Ethical Digital Marketing

1. **Protect Patient Data**
   Follow HIPAA Privacy and Security Rules. Encrypt communications, obtain written consent, and avoid using PHI without authorization.
2. **Verify Medical Claims**
   Ensure that all health-related claims are supported by peer-reviewed evidence or clinical trial data. Avoid unverified outcomes or testimonials.
3. **Disclose Conflicts of Interest**
   Be transparent about sponsorships, product endorsements, and affiliate links.
4. **Audit AI Outputs**
   Review content generated by AI systems for clinical accuracy and unintended bias. Avoid using predictive tools that infer sensitive diagnoses without explicit consent.
5. **Stay Updated on New Laws**
   Track changes to FTC advertising guidelines, HIPAA enforcement, and state privacy laws (e.g., CCPA). Apply GDPR standards when marketing internationally.
6. **Train Your Team**
   Provide regular compliance training for marketing and communications staff. Include case examples, regulatory updates, and AI ethics modules.

Looking ahead, marketers must also prepare for the new ethical challenges that emerging technologies will bring. To support prelaunch decisions, see Appendix 11.5: Ethical Review Questions Before Launching a Campaign, which includes a ten-question checklist to evaluate consent, risk disclosure, tool compliance, and AI content review before a campaign goes live.

Future Trends in Ethical Digital Marketing

Digital healthcare marketing is entering a new era. As technologies such as AI, biometric sensors, and predictive analytics mature, the ethical expectations surrounding data use, patient autonomy, and digital communication are rising. These shifts are driven by both evolving patient expectations and global scrutiny of how personal health data is collected and utilized.

Transparency and Consent

Patients are asking more pointed questions about how their data is used. A 2023 Pew Research Center survey found that 79% of US adults were concerned about how companies use their health information, and 68% wanted the right to approve each use of their data [25]. In response, healthcare marketers will need to implement more granular consent protocols and give patients more control over how and when marketing messages are delivered.

Personalization, Surveillance, and Predictive Risk

New tools allow marketers to customize campaigns using real-time behavioral data, geolocation, wearable device feeds, and even inferred emotional states. Predictive analytics is also being used to target individuals deemed "at risk" for certain diseases before diagnosis. While these tools increase engagement, they also raise ethical concerns about surveillance and bias. For instance, if a digital platform infers a mental health condition and tailors ads without disclosure, the risk of unintended harm grows. Healthcare marketers will need to audit these technologies for fairness, explainability, and nondiscrimination.

Security Pressures and Escalating Expectations

Data breaches remain a serious risk. Sophisticated phishing attacks and AI-driven malware now target healthcare systems with unprecedented precision. Patients expect institutions to respond with advanced protections, including end-to-end encryption, zero-trust frameworks, and formal accountability for third-party vendors. Marketers who handle patient data will be held to the same technical standards as clinicians and IT teams.

Regulatory Evolution

Regulators are moving to address these risks. In 2023, the Federal Trade Commission warned health tech companies that deceptive claims about AI-generated medical advice or health outcomes would face enforcement action [26]. The Department of Health and Human Services has released proposed rule changes to HIPAA that address digital disclosures and third-party tracking tools. Internationally, the European Union's AI Act is expected to restrict how sensitive health data can be processed in marketing algorithms, particularly those that influence patient behavior [27].

How to Prepare for What Comes Next

Healthcare marketers should start by auditing current campaigns for transparency, fairness, and privacy risks. Training programs need to include real-world case examples and prepare teams to meet the requirements of multiple legal frameworks. Collaboration with clinical, legal, and data science teams helps ensure that marketing tools are safe, explainable, and respectful of patient rights. These actions create the foundation for embedding ethics and compliance into daily operations.

Patients are paying attention, and regulators are increasing scrutiny. The future of ethical digital marketing will belong to organizations that act early, plan thoroughly, and keep patient interests at the center of every campaign.

Conclusion: Ethical Marketing Starts with Accountability

Ethical digital marketing in healthcare depends on more than compliance. It requires leadership, sound judgment, and a commitment to respecting patients as people, not targets. This chapter explored the key ethical challenges of digital marketing—from protecting privacy to using AI responsibly—and outlined strategies that align with patient expectations and regulatory standards.

Privacy and data security are critical, with strict adherence to HIPAA guidelines being essential for protecting patient information. The ethical use of AI in healthcare marketing also requires attention, ensuring that these technologies are applied with accuracy, transparency, and fairness. For a concise summary of HIPAA's rules on healthcare marketing, including consent requirements, vendor compliance, and examples of common violations, see Appendix 11.1 HIPAA and Marketing: Quick Guide.

To move forward, marketers must do more than understand the rules; they must design processes that prevent violations, invite accountability, and center the patient in every message. Whether writing ad copy, automating outreach, or analyzing engagement data, the goal should be the same: to use digital tools to support patient care, not exploit it.

What Marketers Should Do Next

- Review all campaigns that use patient data and confirm consent has been obtained and logged.
- Audit AI tools and third-party platforms for accuracy, bias, and HIPAA compliance.
- Train teams regularly on HIPAA, FTC, and emerging AI regulations.
- Create a written code of marketing ethics that includes boundaries for personalization.
- Implement feedback channels so patients can report concerns about digital outreach.

These are not one-time actions. They should become part of routine marketing operations, with clear ownership and regular review. Table 11.1 summarizes five core ethical principles in healthcare marketing with plain-language descriptions.

Summary Table: Five Core Ethical Principles in Healthcare Marketing

Principle	Description
Transparency	Disclose how patient data is collected, stored, and used
Consent	Obtain clear, documented approval before using health data in campaigns
Privacy	Limit data collection and store information securely
Truthfulness	Ensure that all claims are evidence-based and not misleading
Fair use of AI	Audit algorithms for accuracy, fairness, and unintended harm

By building these principles into every campaign, healthcare marketers can create digital strategies that protect patients, enhance trust, and set a standard others will follow.

Appendices

Appendix 11.1

HIPAA and Marketing: Quick Guide

How to Use This Guide

This appendix provides a practical reference for healthcare professionals and marketers who need to comply with HIPAA regulations while promoting services. It outlines what qualifies as protected health information (PHI), when patient consent is required, and when Business Associate Agreements (BAAs) must be in place. Use this guide to avoid common pitfalls in digital outreach, evaluate vendor compliance, and ensure that your marketing practices protect patient privacy and follow federal law.

What Counts as Protected Health Information (PHI)

PHI includes any individually identifiable health information in any form, oral, paper, or electronic, that relates to the following:

- Past, present, or future physical or mental health conditions.
- Provision of healthcare.
- Payment for healthcare.
- Common identifiers such as name, address, birthdate, Social Security number, email, or IP address linked to health data.

Marketing that Requires Patient Authorization

Under HIPAA, marketing is any communication that encourages a patient to purchase or use a product or service, **unless**

- It is a face-to-face communication by the provider.
- It involves a promotional gift of nominal value.
- It relates to a patient's treatment or care coordination, *and* the provider does not receive financial payment from a third party.

Authorization Is Required If

- The communication promotes third-party products or services.
- A vendor pays the healthcare organization to send the message.
- Patient data is being used to target or personalize advertising outside of treatment or operations.

Business Associate Agreements (BAAs) Are Required When

You share PHI with any third party that handles data on your behalf, including the following:

- Email marketing platforms (e.g., Mailchimp, Constant Contact).
- Customer relationship management (CRM) systems.
- Advertising platforms with patient targeting capabilities.
- Data analytics or reputation management vendors.

Examples of Common HIPAA Violations in Marketing

- Including patient testimonials without written authorization.
- Using tracking pixels on patient-facing websites without a BAA.

- Sending unencrypted emails with appointment reminders that contain health details.
- Sharing patient email addresses with third-party marketers without consent.

Best Practices

- Obtain written consent before using PHI in marketing.
- Use only HIPAA-compliant platforms for email or text outreach.
- Review vendor contracts and ensure a signed BAA is in place.
- Avoid using cookies or tracking technologies without HIPAA-compliant configurations.

Appendix 11.2

Sample Consent Language for Digital Platforms

Purpose of This Appendix

This appendix provides a sample of legally appropriate, plain-language consent text for healthcare organizations using digital platforms to communicate with patients. The goal is to help you meet HIPAA and marketing compliance standards while ensuring that patients understand what they are agreeing to. This template can be used for forms that involve appointment scheduling, newsletter sign-ups, digital reminders, or other patient communications involving personal information.

Sample Consent Language

By providing your contact information, you agree to receive communications from [Practice Name] regarding your care, appointment reminders, and health-related information. These messages may include newsletters, updates, or service announcements. We will never sell your information. Communications may be sent by email, text message, or phone call. You can opt out at any time.

We may use secure third-party services to manage communications. These services are required by law to protect your privacy and follow HIPAA rules. For more information about how we protect your data, please review our Privacy Policy.

☐	*I agree to receive digital communications as described above.*
☐	*I have read and understand the privacy practices of [Practice Name].*

Use Guidelines

- Always pair this language with a link to your full privacy policy.
- Only use third-party platforms with a signed Business Associate Agreement (BAA).
- Offer opt-out or unsubscribe options in every communication.
- Modify "[Practice Name]" to reflect your organization's name.

Appendix 11.3

Ethical Marketing Red Flags Checklist

Use this checklist to evaluate digital campaigns before they go live. If you check any of these boxes, pause and reassess your content for compliance and integrity.

☐ Red Flag	☐ Why It's a Problem
☐ Claims a guaranteed outcome	Misleads patients; violates FTC standards
☐ Uses patient photos or quotes without signed consent	Violates HIPAA and patient privacy rights
☐ Hides or omits risks or side effects	Undermines informed decision-making
☐ Deletes or suppresses negative reviews	Misrepresents real patient experience
☐ Fails to disclose business ties or endorsements	Breaches transparency requirements
☐ Uses superlatives like "#1 rated" without proof	Can be considered deceptive marketing
☐ Promotes unapproved treatments without clarification	May violate FDA regulations
☐ Employs fear, guilt, or urgency to drive action	Exploits vulnerable audiences

How to Use This Checklist

- Review all digital materials before publication.
- Use it during internal marketing or ethics committee reviews.
- If any item is marked, revise the content or seek legal/ethical consultation.

Appendix 11.4

Recommended HIPAA-Compliant Tools for Marketing

These platforms support digital marketing while meeting HIPAA privacy and security requirements. Confirm that a signed Business Associate Agreement (BAA) is in place before use.

Category	Tool	Description
Email marketing	Paubox	End-to-end encrypted email. No portals or extra log-ins for patients
	LuxSci	Secure email with marketing automation features
	Virtru	Gmail-compatible encryption with compliance tracking
SMS/text messaging	OhMD	HIPAA-compliant texting, appointment reminders, and two-way messaging
	TigerConnect	Secure messaging platform used by hospitals and clinics
	Spruce health	Combines secure messaging, telehealth, and practice management
Analytics	Freshpaint	Tracks user behavior without sending PHI to third parties. Offers consent tools
	Segment (with restrictions)	Can be HIPAA-compliant with proper configurations and BAAs
	Datavant	Used for de-identified data aggregation and analytics
CRM platforms	Salesforce health cloud	Patient relationship management with built-in HIPAA compliance
	HubSpot (Enterprise tier)	HIPAA compliance possible with custom setup and signed BAA
	NexHealth	Syncs with EHRs for secure patient outreach and scheduling

Use Guidelines

- Always request and retain a signed BAA from vendors.
- Do not store or transmit PHI using tools that lack formal HIPAA support.
- Use platform features like audit logs, access control, and encryption.
- Pair tools with internal policies on consent, access, and data handling.

Appendix 11.5

Ethical Review Questions Before Launching a Campaign

Use this ten-question tool as part of your final review process before publishing any digital marketing campaign. If any answer is "No" or "Unsure," pause the launch and investigate further.

Yes/ No	Review question
☐	Have we obtained written, documented consent for any patient quotes, photos, or stories used?
☐	Does the content avoid making unsubstantiated claims, such as "guaranteed results"?
☐	Have we clearly disclosed any risks, limitations, or side effects of the services promoted?
☐	Is all content free from emotionally manipulative language like fear, urgency, or guilt?
☐	Have we confirmed that no protected health information (PHI) is exposed without authorization?
☐	Are all third-party tools used in this campaign HIPAA-compliant and covered by a BAA?
☐	Has an internal compliance, legal, or ethics review been completed for this campaign?
☐	Have we disclosed any financial interests, endorsements, or affiliate partnerships?
☐	Are all AI-generated materials in this campaign reviewed for bias, accuracy, and relevance?
☐	Have we provided patients a way to opt out of communications or data collection?

How to Use

- Integrate this checklist into your final prelaunch workflow.
- Use it for every campaign involving patient engagement, digital targeting, or AI tools.
- Require at least two team members to review and sign off on the checklist for accountability.

References

1. Guardian.Nine out of ten beauty clinics breaking the law by advertising Botox. The Guardian. 2023. https://www.theguardian.com/media/2023/may/07/advertising-watchdog-cracks-down-on-misleading-health-and-beauty-claims?utm_source=chatgpt.com. Accessed 2 Aug 2025.
2. Kliff S, Lecher C. Facebook is receiving sensitive medical information from hospital websites. The Markup. 2022. https://themarkup.org/pixel-hunt/2022/06/16/facebook-is-receiving-sensitive-medical-information-from-hospital-websites. Accessed 2 Aug 2025.
3. McCoy TH, Perlis RH, Adida B. Hospitals' use of third-party tracking code on their websites. Health Aff. 2022;41(12):1735–43. https://doi.org/10.1377/hlthaff.2022.01119.

4. Komenda, E. Hospitals sued over online tracking tools that may violate patient privacy. USA TODAY. 2023. https://www.usatoday.com/story/news/health/2023/04/13/hospital-lawsuits-meta-pixel-violate-patient-privacy/11665083002/.
5. Auxier B, Rainie L, Anderson M, Perrin A, Kumar M, Turner E. Americans and privacy: concerned, confused and feeling lack of control over their personal information. Pew Research Center; 2019. https://www.pewresearch.org/internet/2019/11/15/americans-and-privacy-concerned-confused-and-feeling-lack-of-control-over-their-personal-information/.
6. U.S. Department of Health & Human Services. Summary of the HIPAA privacy rule. 2025. hhs.gov+1hhs.gov+1.
7. Alder, S. One-third of healthcare websites still use meta pixel tracking code. HIPAA J. 2024. https://www.hipaajournal.com/one-thirdhealthcare-websites-meta-pixel-tracking-code-2024/.
8. U.S. Department of Health & Human Services. 2025. Summary of the HIPAA privacy rule. hhs.gov+1hhs.gov+1.
9. Rivelli, E. Do business associates need to comply with HIPAA? Insureon. 2025. https://www.insureon.com/blog/do-business-associates-needhipaa-complianceinsureon.com.
10. U.S. Department of Health & Human Services. Direct Liability of Business Associates. 2021. https://www.hhs.gov/hipaa/forprofessionals/privacy/guidance/business-associates/factsheet/index.html?utm_source=chatgpt.com.
11. American Dental Association (ADA). What are the penalties for violating HIPAA? 2025. https://www.ada.org/resources/practice/legal-andregulatory/hipaa/penalties-for-violating-hipaa?utm_source=chatgpt.com.
12. Mello MM, McCoy MS. Health data privacy and consent in the digital era. N Engl J Med. 2020;382(10):975–8. https://doi.org/10.1056/NEJMp1917160.
13. Baker A, Meslin EM, Yu J. Measuring comprehension in digital consent forms: an experimental comparison. J Med Ethics. 2021;47(2):122–8. https://doi.org/10.1136/medethics-2019-105962.
14. Children's Hospital Association. "Made Possible by Your Children's Hospital" campaign wins award for best visual storytelling. CHA News. 2024. https://www.childrenshospitals.org/news/newsroom/2024/10/made-possible-campaign-wins-prnews-award?utm_source=chatgpt.com.
15. HIMSS. 2021 HIMSS Healthcare Cybersecurity Survey. 2021. https://www.himss.org/resources/2021-himss-healthcare-cybersecurity-survey.
16. Ponemon Institute. The impact of ransomware on healthcare during COVID-19 and beyond. Sponsored by Censinet. 2022. https://www.censinet.com/ponemon-report-healthcare-ransomware-2022/.
17. National Institute of Standards and Technology. Security and Privacy Controls for Information Systems and Organizations (NIST SP 800–53 Rev. 5). 2020. https://doi.org/10.6028/NIST.SP.800-53r5.
18. HITRUST Alliance. HITRUST CSF v9.6.2 Framework Overview. 2021. https://hitrustalliance.net/hitrust-csf/.
19. Lopez G. This company used smartphone data to target ads to opioid addicts. Vox. 2018. https://www.vox.com/policy-and-politics/2018/12/30/18159859/opioid-addiction-targeted-advertising-ethics.
20. Salesforce. How memorial Sloan Kettering cancer center enhances patient experiences with salesforce health cloud. 2022. Retrieved from https://www.salesforce.com/resources/articles/mskcc-patient-engagement/.
21. U.S. Department of Health and Human Services. HIPAA privacy rule and marketing. 2022. https://www.hhs.gov/hipaa/for-professionals/privacy/guidance/marketing/index.html.
22. HHS Office for Civil Rights. Resolution agreement with Phoenix healthcare group. 2023. https://www.hhs.gov/hipaa/for-professionals/compliance-enforcement/agreements/index.html.
23. Federal Trade Commission. Advertising and marketing basics for professionals. 2023. https://www.ftc.gov/business-guidance/resources/advertising-marketing-basics.

24. Federal Trade Commission. AI and health marketing: proceed with caution. 2023. https://www.ftc.gov/news-events/blogs/business-blog/2023/03/ai-health-marketing-proceed-caution.
25. Pew Research Center. Americans' views about privacy, Surveillance and data collection. 2023. https://www.pewresearch.org/internet/2023/06/15/americans-views-about-privacy-surveillance-and-data-collection.
26. Federal Trade Commission. FTC warns health apps and connected devices to comply with health breach notification rule. 2023. https://www.ftc.gov/news-events/news/press-releases/2023/01/ftc-warns-health-apps-connected-devices-comply-health-breach-notification-rule.
27. European Commission. Proposal for a regulation on artificial intelligence (AI act). 2023. https://digital-strategy.ec.europa.eu/en/policies/european-approach-artificial-intelligence.

Chapter 12
Feedback and Peer Review Mechanisms in Digital Marketing for Healthcare

Introduction

In healthcare, digital marketing is a key channel for connecting with patients and improving the visibility of services. Feedback and peer review are essential to making these efforts accurate, effective, and trustworthy.

Feedback provides direct insight from patients and healthcare consumers about their online experiences, preferences, and perceptions of digital campaigns. This information helps marketers adjust messaging, improve usability, and address gaps in communication. For example, a hospital that uses patient survey results to simplify its appointment booking page may see both higher satisfaction scores and increased appointment completion rates.

Peer review brings a professional perspective to content evaluation. Involving clinicians, compliance officers, and marketing specialists ensures that digital materials are factually correct, ethically sound, and consistent with brand standards. A clinic that implemented peer review for its social media campaigns reduced regulatory risk by removing unsupported treatment claims before publication.

Together, feedback and peer review create a continuous improvement loop that aligns marketing with patient needs, regulatory requirements, and professional ethics.

C. Quinn, *The Digital Practice*, https://doi.org/10.1007/978-3-032-07745-5_12

Sidebar: Ethical Peer Review in Practice

Use this checklist to evaluate healthcare marketing content before it goes live:

☐	**Verify medical accuracy and currency:** Ensure every clinical claim reflects the latest peer-reviewed evidence or established guidelines.
☐	**Avoid exaggerated or misleading claims:** Keep statements balanced and avoid superlatives that suggest guaranteed outcomes.
☐	**Include qualified reviewers:** At least one reviewer should have clinical training, and another should have expertise in healthcare marketing regulations or institutional compliance.
☐	**Check alignment with privacy, brand, and regulatory standards:** Confirm content meets HIPAA requirements, your organization's branding, and any relevant FTC or FDA advertising rules.
☐	**Ensure respectful, inclusive language:** Avoid stigmatizing terms and write for audiences with varied health literacy levels.

Tip: Document the peer review process by noting who reviewed the content, when the review occurred, and what changes were made. This record supports accountability and can be valuable in regulatory audits.

Gathering Feedback on Digital Marketing Efforts

Collecting and applying patient feedback is essential for improving both healthcare services and marketing strategies. Digital platforms make this process more efficient through surveys, feedback forms, and social media interactions. The chosen platform should match the habits and preferences of the target audience to maximize response rates and the usefulness of the data.

Surveys and Feedback Forms

Online surveys allow the structured collection of information about patient experiences and satisfaction. They can be tailored to focus on specific services or communication channels. HIPAA compliance is critical when these tools capture or store protected health information (PHI).

Table 12.1 compares common survey tools, HIPAA status, best uses, and limits.

Social Media Feedback

Platforms like Facebook, Instagram, and Twitter/X offer real-time interaction with patients and the community. These channels are valuable for understanding public sentiment and answering questions, but they carry privacy risks. Healthcare

Table 12.1 Survey tools for healthcare marketing, HIPAA status, best uses, and limits

Tool	HIPAA-compliant	Best for	Limitations
Google forms	No	Basic feedback with low privacy risk; suitable for non-identifiable responses	No encryption or audit logs; unsuitable for PHI unless under a rare, covered entity agreement[a]
SurveyMonkey	Enterprise version only (with signed BAA)	Patient satisfaction surveys, net promoter score tracking, post-visit follow-ups	Standard plans lack HIPAA compliance; BAA only available with Enterprise tier
Qualtrics	Yes (business and health editions)	Advanced research, experience management, detailed analytics	Higher cost; complex interface requiring training
REDCap	Yes	Longitudinal clinical research, quality improvement projects, secure institutional use	Institutional access required; not user-friendly for marketing purposes
Jotform	Yes (with signed BAA)	Intake forms, basic feedback, appointment requests	Risk of misconfiguration without administrator oversight

Tip: Always confirm whether your plan includes a signed Business Associate Agreement (BAA) and whether data is encrypted at rest and in transit

[a]A “rare, covered entity agreement” refers to a custom Business Associate Agreement (BAA) with Google that specifically includes Google Forms. This arrangement is uncommon, and without it, Google Forms should only be used for non-identifiable feedback

organizations must avoid collecting PHI in public threads or direct messages without secure systems in place.

For instance, the Memorial Hermann Health System in Texas uses Qualtrics to collect real-time patient feedback via post-visit surveys and onsite QR codes. This strategy revealed service gaps such as meal delivery delays and confusion about signage. They deployed targeted interventions and immediately improved patient satisfaction, achieving an 8.2% Net Promoter Score increase in outpatient areas and over a ten-point rise in emergency department NPS during concierge shifts.,[1,2]

Analytics for Evaluation

Feedback mechanisms are most valuable when paired with analytics tools such as Google Analytics or Hootsuite. These track engagement metrics including page visits, survey completion rates, and comment sentiment. Metrics such as bounce rate,

[1] https://www.qualtrics.com/experience-management/industry/patient-experience/?utm_source=chatgpt.com. Accessed 8/9/25.

[2] https://www.qualtrics.com/customers/memorial-hermann/?utm_source=chatgpt.com. Accessed 8/8/25.

average time on page, and conversion rate help marketers evaluate whether digital campaigns are prompting desired actions.

By selecting secure, user-friendly feedback tools and analyzing results systematically, healthcare marketers can make informed changes that improve both patient experience and campaign effectiveness.

While patient feedback captures the patient's perspective, peer review adds a professional lens, ensuring that marketing content is accurate, ethical, and aligned with both clinical standards and organizational goals. Alongside patient feedback, a structured peer review process adds professional oversight, helping ensure that digital marketing content meets both clinical and regulatory standards.

Implementing Peer Review Processes

Establishing a robust peer review process among healthcare professionals is a critical step toward enhancing the effectiveness of digital marketing strategies within the healthcare sector. The process begins with the careful selection of participants, ensuring a diverse representation of expertise and perspectives. Participants are typically chosen based on their experience, knowledge in digital marketing, and role in healthcare delivery. The criteria for evaluation are clearly defined, focusing on the relevance, coherence, effectiveness, efficiency, and impact of the digital marketing strategies employed.[3] Reviews are conducted at regular intervals, allowing for the continuous assessment and improvement of marketing efforts.

The impact of peer reviews on digital marketing outcomes in healthcare can be significant. For example, a study by Bader [1] highlighted how peer review processes led to the identification of best practices and areas for improvement in patient care delivery, which can be directly applied to digital marketing strategies [1]. Another case studied involved a healthcare facility that implemented peer reviews to evaluate its social media strategies, resulting in increased patient engagement and satisfaction due to more targeted and relevant content.

The peer review process provides a structured approach for healthcare professionals to critically evaluate and enhance their digital marketing strategies. By focusing on the selection of knowledgeable participants, setting clear evaluation criteria, and maintaining a regular review schedule, healthcare organizations can leverage peer reviews to achieve better marketing outcomes, ultimately improving patient care and engagement.

Peer review does more than improve operations. In digital marketing, it helps refine the tone and clarity of outreach materials, ensures consistency with brand values, and flags content that may misrepresent services or violate regulatory standards.

---

[3] https://www.oecd.org/dac/evaluation/daccriteriaforevaluatingdevelopmentassistance.htm. Accessed 3/21/24.

Once feedback from patients and peers has been collected, the next step is to analyze it in a structured way so it can guide specific changes to marketing strategies.

Analyzing Feedback for Actionable Insights

Analyzing feedback and peer review data is essential for refining digital marketing strategies in healthcare.[4] This process involves both quantitative and qualitative analysis to uncover actionable insights. Quantitative data, such as survey scores and completion rates, can be analyzed using statistical tools to identify trends and patterns.

For example, a consistent drop in survey scores related to patient wait times could indicate a need for operational adjustments.[5] Qualitative feedback, such as comments and suggestions received through feedback forms or social media, requires content analysis to understand the sentiment and specific concerns of patients. The value of analysis depends on the quality of the feedback received, which makes it essential to design processes that encourage active and honest participation from both patients and peers.

The following flowchart (Fig. 12.1) illustrates how patient feedback and peer review work together to refine digital marketing strategies in healthcare.

Thematic Analysis

Techniques like thematic analysis can also be employed. Thematic analysis in the context of healthcare digital marketing is a qualitative research method that focuses on identifying, analyzing, and reporting patterns (themes) within data. This approach is particularly useful for understanding the complex experiences, behaviors, and perceptions of patients and healthcare consumers as they interact with digital marketing content.[6]

In healthcare digital marketing, thematic analysis can be applied to various forms of feedback and data, such as social media comments, survey responses, and online reviews [2]. By examining these data sources, marketers can uncover recurring themes that may indicate patients' preferences, concerns, and expectations regarding healthcare services and information [3]. For example, a thematic analysis of patient feedback on a hospital's social media page might reveal themes related to the desire for more informative content about specific treatments or appreciation for prompt responses to inquiries.

---

[4] https://qualaroo.com/blog/customer-feedback-analysis/. Accessed 3/21/24.

[5] https://userpilot.com/blog/feedback-analysis/. Accessed 3/21/24.

[6] https://www.sciencedirect.com/topics/social-sciences/thematic-analysis. Accessed 3/21/24.

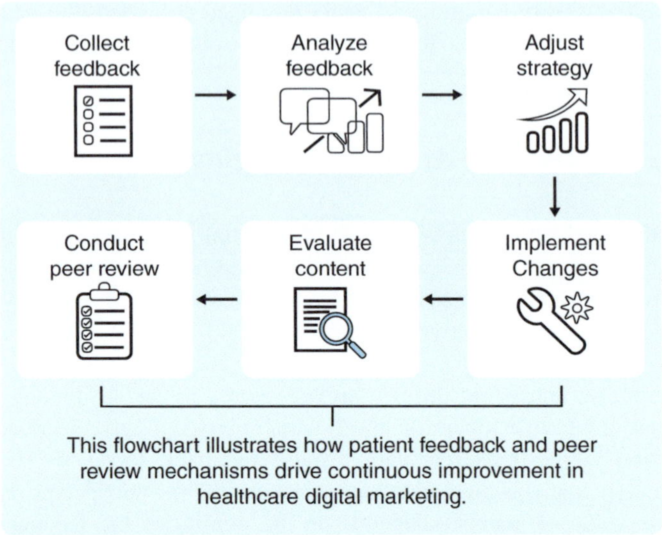

Fig. 12.1 How feedback and peer review improve healthcare marketing. (Image by Quinn/ChatGPT)

Several healthcare organizations have leveraged advanced analysis methods to turn feedback into operational insight. Researchers at the University of Oxford analyzed hundreds of thousands of free-text patient comments across NHS trusts, using thematic analysis to uncover recurring concerns, such as communication and wait times, and directly inform quality improvement initiatives [4].

In the United States, Penn State collaborators used machine learning models on Geisinger patient satisfaction data [5]. Their AI identified key themes like staff courtesy and pain management, which helped health system leaders target improvements in messaging and service delivery.

Sidebar: What Is Thematic Analysis

What It Is

Thematic analysis is a way of examining written or spoken feedback to find common ideas or topics, called "themes." It is often used with open-ended answers from surveys, online reviews, or patient interviews.

Why It Matters in Healthcare Marketing

It helps healthcare organizations understand what patients talk about most, what they value, and what problems they experience. This information can guide improvements in communication, service delivery, and marketing messages. Unlike sentiment analysis, which measures the emotional tone of feedback, thematic analysis focuses on identifying recurring topics or themes.

(continued)

Common Use Cases

- **Survey comments:** Group patient feedback into topics like wait times, staff courtesy, or clarity of instructions.
- **Online reviews:** Identify trends in patient attitudes, such as trust, confusion, or dissatisfaction, that can influence the tone of future campaigns.
- **Patient interviews or focus groups:** Turn in-depth conversations into clear, actionable ideas for service changes or new marketing approaches.

How It Works

1. Read through all the responses and highlight repeated words or ideas.
2. Combine similar ideas into larger topic groups.
3. Give each group a short, clear label (e.g., "access," "professionalism," or "billing concerns").
4. Use these themes to adjust services, create new content, or improve patient communication.

Caution

To avoid misinterpretation, have more than one person review the results, and check them against clinical or operational feedback. This helps ensure that the conclusions are accurate and balanced.

Incorporating this feedback into marketing strategies involves identifying the most significant areas for improvement and aligning marketing messages with patient needs and preferences. For instance, if feedback highlights a lack of understanding about a particular service, marketing content can be adjusted to include more educational information about that service.

Additionally, positive feedback themes can be amplified in marketing materials to highlight strengths and build trust with potential patients. An example of this could be leveraging patient testimonials about positive experiences with telehealth services to encourage more patients to try this option [6].

By systematically analyzing feedback and making informed adjustments to marketing strategies, healthcare providers can enhance patient engagement, improve service delivery, and ultimately foster a more patient-centric approach to care.

Best Practices for Engaging Patients and Peers in the Feedback Process

To foster active participation in the feedback process from both patients and healthcare professionals, it's essential to create an environment that encourages open and honest communication. One effective strategy is to ensure that feedback mechanisms are easily accessible and user-friendly. Digital platforms like online surveys, feedback forms, and social media can facilitate this, but it's crucial to choose the right tools that resonate with the target audience [7].

As outlined in the earlier survey tool comparison table, each platform offers different technical features and compliance options. In practice, the most effective choice depends on how well it encourages participation and makes the process easy for patients and peers.

- **Ease of access:** Tools that work on mobile devices and require minimal steps, such as HIPAA-compliant Jotform forms embedded on a portal, make it more likely that patients will complete them.
- **Real-time interaction:** Social media platforms like Facebook or Instagram can boost engagement when used for quick polls or live Q&A sessions, provided PHI is not collected in public threads.
- **Patient motivation:** Survey tools with personalized thank-you messages or visible progress bars, such as Qualtrics, can help maintain participation rates and show patients their input is valued.

Determining the right tools involves analyzing your audience's demographics, preferences, and digital behaviors. Surveys can provide in-depth insights but require a commitment to complete, making them suitable for engaged patients. Feedback forms are less time-consuming and can capture quick impressions, ideal for general feedback. Social media interactions offer real-time engagement but may require active management and moderation.

Ultimately, the choice of tool should align with your audience's habits and preferences, ensuring that the feedback process is as effortless as possible for them, thereby increasing the likelihood of their participation.

Creating a culture of trust and transparency within healthcare settings can significantly enhance the willingness of patients and professionals to share their insights. This involves clearly communicating the purpose of collecting feedback and how it will be used to improve services. Additionally, recognizing and appreciating the time and effort participants put into providing feedback can further motivate engagement.[7]

Addressing ethical and privacy concerns is paramount when collecting and utilizing feedback. Healthcare providers must adhere to laws and regulations regarding patient confidentiality, such as HIPAA in the United States. This includes obtaining informed consent for the use of patient data, limiting access to feedback to authorized personnel only, and employing secure methods for data collection and storage [8]. Ensuring that feedback remains anonymous can help mitigate privacy concerns and encourage more candid responses.

By implementing these strategies, healthcare providers can enhance the feedback process, leading to valuable insights that drive improvements in patient care and service delivery.

When feedback is gathered consistently and ethically, it becomes a powerful tool for ongoing improvement and helps marketing strategies evolve alongside patient needs and market conditions.

[7] https://implementationscience.biomedcentral.com/articles/10.1186/s13012-018-0784-z. Accessed 3/21/24.

Leveraging Feedback for Continuous Improvement

In the dynamic landscape of healthcare digital marketing, feedback and peer reviews are invaluable for fostering continuous improvement and ensuring that marketing strategies are closely aligned with patient needs and market dynamics. This iterative approach to strategy development allows healthcare organizations to remain agile, responsive, and patient-centered in their marketing efforts.

Iterative strategy development: Feedback and peer reviews act as critical inputs for the iterative development of marketing strategies. By systematically collecting and analyzing feedback from patients and healthcare professionals, organizations can identify areas of strength and opportunities for enhancement.

For example, a healthcare provider might use patient feedback collected through online surveys to understand patient perceptions of their telehealth services. If patients express concerns about the ease of use of the telehealth platform, the provider can then prioritize improvements to the user interface and user experience in their marketing and service delivery strategies. This process of continuous feedback, analysis, and adjustment ensures that marketing strategies evolve in step with patient expectations and technological advancements.[8]

Leveraging feedback and peer reviews is not only used to refine marketing messages and tactics but also to enhance the overall patient experience. By embracing an iterative approach to strategy development, healthcare organizations can ensure their marketing efforts are both effective and deeply rooted in the needs and preferences of their patients.

Looking ahead, advances in technology are reshaping how healthcare organizations collect, analyze, and act on feedback, making the process faster, more precise, and more patient-centered.

Future Trends in Feedback and Peer Review Mechanisms

The future of feedback and peer review mechanisms in healthcare digital marketing is poised for significant evolution, driven by technological advancements and the integration of new tools and platforms. As digital marketing continues to evolve, so too will the methods by which healthcare organizations collect and analyze feedback from patients and peers.

Predictions for evolution: Advancements in artificial intelligence (AI) and machine learning (ML) are expected to play a pivotal role in transforming feedback and peer review processes. These technologies can automate the analysis of large volumes of feedback data, identifying patterns and insights more quickly and accurately than manual methods.

---

[8] https://fastercapital.com/topics/leveraging-user-feedback-for-continuous-improvement-in-user-experience-design.html. Accessed 3/21/24.

For example, natural language processing (NLP) algorithms can interpret open-ended responses in surveys, extracting meaningful themes and sentiments without human bias. This capability will enable healthcare marketers to gain deeper insights into patient experiences and preferences, facilitating more personalized and effective marketing strategies.

Preparation for emerging tools: To capitalize on these advancements, healthcare organizations must prepare for the integration of new tools and platforms that enhance feedback collection and analysis. This preparation involves investing in AI and ML technologies and training staff to utilize these tools effectively. Additionally, healthcare marketers should explore platforms that offer real-time feedback capabilities, such as social listening tools, which monitor social media and online forums for mentions of their organization or services. By adopting these technologies, healthcare providers can stay ahead of patient needs and expectations, adapting their marketing strategies in real time based on direct patient feedback.

However, these technologies carry important risks. AI-driven analysis can produce biased results if the training data is unrepresentative, and automated sentiment scoring may misinterpret patient tone or context. Without ongoing human oversight and validation, these errors can lead to misleading conclusions, flawed marketing strategies, and a loss of patient trust.

The future of feedback and peer review mechanisms in healthcare digital marketing is set to become more dynamic, precise, and patient-focused. By embracing technological advancements and preparing for the integration of emerging tools, healthcare organizations can ensure their marketing efforts are both responsive and aligned with the evolving landscape of patient care and communication.

Documenting Changes for Accountability

Keeping a record of the actions taken in response to feedback builds transparency and trust. This documentation also serves as a reference for future decisions, helping teams avoid repeating past mistakes and track the effectiveness of implemented changes.

Key Steps in Iterative Improvement

- Review and prioritize feedback based on impact and feasibility.
- Plan and implement targeted changes.
- Communicate updates to stakeholders and, when appropriate, to patients.
- Monitor outcomes to ensure improvements are effective.
- Reassess and adjust based on new feedback.

By treating feedback as part of an ongoing improvement cycle and documenting results, organizations can close the loop and build a stronger foundation for future strategies.

Conclusion

Feedback and peer review are essential to effective healthcare digital marketing. These processes help marketers align strategies with patient expectations, identify areas for improvement, and build trust through transparency and responsiveness.

When organizations create systems that welcome feedback and promote internal review, they improve both communication and campaign outcomes. Peer input can catch errors, sharpen messaging, and ensure compliance. Patient insights reveal gaps in understanding, usability, or emotional tone, allowing teams to adjust content in real time.

Digital marketing in healthcare demands flexibility. Preferences shift, platforms evolve, and patient expectations grow. A feedback-driven approach enables organizations to keep pace. Those that respond quickly and thoughtfully stand out, strengthen engagement, and improve patient satisfaction.

In a crowded digital landscape, continuous review isn't optional. It's the only path to relevance, resilience, and lasting connection.

References

1. Bader H, Abdulelah M, Maghnam R, Chin D. Clinical peer review; a mandatory process with potential inherent bias in desperate need of reform. J Community Hosp Intern Med Perspect. 2021;11(6):817–20. https://doi.org/10.1080/20009666.2021.1965704. PMID: 34804397; PMCID: PMC8604442. Accessed 21 Mar 2024.
2. Braun V, Clarke V. Using thematic analysis in psychology. Qual Res Psychol. 2006;3(2):77–101.
3. Kordzadeh N, Young DK. Exploring hospitals' use of Facebook: thematic analysis. J Med Internet Res. 2018;20(5):e190. https://doi.org/10.2196/jmir.9549. PMID: 29769173; PMCID: PMC5976865. Accessed 21 Mar 2024.
4. Saleem, S. Transforming patient experience through thematic analysis of NHS feedback. Presentation at the International Healthcare Conference. 2024. Retrieved from https://4149027.fs1.hubspotusercontent-na1.net/hubfs/4149027/IH24%20Speaker%20presentations/Sophia%20Saleem.pdf.
5. Penn State News. Penn State researchers use AI to analyze patient satisfaction. Penn State University. 2023. Retrieved from https://www.psu.edu/news/research/story/penn-state-researchers-use-ai-analyze-patient-satisfaction.
6. Purcarea VL. The impact of marketing strategies in healthcare systems. J Med Life. 2019;12(2):93–6. https://doi.org/10.25122/jml-2019-1003. PMID: 31406509; PMCID: PMC6685306
7. Hardavella G, Aamli-Gaagnat A, Saad N, Rousalova I, Sreter KB. How to give and receive feedback effectively. Breathe (Sheff). 2017;13(4):327–33. https://doi.org/10.1183/20734735.009917. PMID: 29209427; PMCID: PMC5709796.
8. Tariq RA, Hackert PB. Patient Confidentiality. [Updated 2023 Jan 23]. In: StatPearls [Internet]. Treasure Island (FL): StatPearls Publishing; 2024. Available from: https://www.ncbi.nlm.nih.gov/books/NBK519540/.

Chapter 13
The Future of Digital Marketing for Medical Practices

Predictions for the Evolution of Digital Marketing in Healthcare

Digital marketing in healthcare will continue to evolve rapidly as patients expect more personalization and providers adopt new technologies. A 2025 Deloitte survey found that nearly 90% of health system executives expect the expanded use of virtual care and connected digital tools to redefine their strategies this year. Health systems also prioritize personalized experiences, with 72% citing improving consumer trust and engagement as a top goal [1].

Telehealth usage, which rose steeply during the COVID-19 pandemic, remains strong. Analysts project virtual care could account for 20–30% of all outpatient encounters in the years ahead [2]. At the same time, wearable devices and remote monitoring are reshaping care delivery [3]. The Boston Consulting Group anticipates smart implants and home diagnostics will drive continuous patient monitoring and remote care transitions in 2025 [4].

Artificial intelligence and data analytics will fuel personalization at scale. Providers will use predictive models across EHR and patient behavior data to anticipate care needs, recommend services, and tailor messaging. Privacy will remain central. As new tools gather more patient data, providers must uphold HIPAA compliance and maintain transparency.

This phase of digital transformation offers healthcare practices new ways to build trust, support patient engagement, and improve outcomes, all while navigating evolving regulations and technology standards.

C. Quinn, *The Digital Practice*, https://doi.org/10.1007/978-3-032-07745-5_13

Personalization at Scale

Digital marketing in healthcare is undergoing rapid transformation as technologies like artificial intelligence (AI) and data analytics reshape how organizations communicate with patients. Personalized marketing, particularly through predictive analytics, patient segmentation, and recommendation engines, is emerging as a core strategy. These tools enable healthcare marketers to develop content aligned with each patient's health status, preferences, and behaviors.

This approach enhances engagement and improves outcomes. For instance, a diabetic patient may receive personalized nutritional tips, medication reminders, and educational videos tailored to their clinical history and lifestyle patterns. Studies have shown that personalization improves medication adherence and patient satisfaction in chronic care populations [5].

When patients receive communications that align with their needs, they are more likely to act on medical advice and stay connected to their provider. This reinforces continuity of care and promotes proactive health behaviors. As the healthcare industry becomes more consumer-driven, integrating AI-powered personalization will be necessary for practices that want to retain patients and stay competitive.

Integration of Emerging Technologies

Virtual reality (VR) and augmented reality (AR) are changing how healthcare providers educate patients and promote their services. These technologies use 3D simulations and real-time visual overlays to explain diagnoses, procedures, and treatment options more clearly than static images or written descriptions.

Providers increasingly embed AR and VR tools into their websites, apps, and waiting room screens to improve digital engagement. For example, clinics may offer virtual walkthroughs of surgical procedures or interactive AR-based animations that explain disease progression. While specific case studies remain limited in the public domain, reviews show that immersive technologies improve patient comprehension and reduce anxiety during care planning.

A 2024 systematic review by Iqbal et al. in *JMIR Medical Education* found that patients using AR and VR tools retained more medical information, experienced less stress, and reported greater satisfaction with their providers [6, 7]. These tools also support post-discharge education and follow-up care, helping patients understand and remember instructions more effectively. Though research on marketing-specific metrics remains limited, this improved comprehension and emotional comfort can translate into longer on-site engagement, stronger brand impressions, and higher patient trust.

The immersive format also holds potential in provider branding and differentiation. Some hospitals use VR to introduce new patients to their facilities and staff.

Others integrate AR into health portals to guide patients through routine processes like medication management or appointment prep. These experiences promote a sense of transparency and ease, which influences patient choice when searching online for care.

Technologies like AR and VR also have the potential to reduce patient anxiety by providing a peaceful and engaging way to learn about medical procedures and health management, as noted by Mutabazi in their discussion on the impact of AR and VR in the healthcare sector.[1]

To use these tools effectively, providers must keep the clinical content accurate, protect patient privacy, and ensure accessibility for users with visual, cognitive, or motor impairments. They should also track outcomes, such as patient activation scores and bounce rates, to evaluate the return on investment. Patient activation scores measure how confident and capable patients feel in managing their health, and higher scores often correlate with better adherence to care plans.

Moreover, the metaverse, as explored by Lewis (2024), provides medical learners with immersive learning experiences in realistic 3D environments. This not only aids in medical education but also in patient understanding of their health conditions [8].

As the adoption of immersive tools increases, so does the need for compliance. Providers must ensure medical accuracy, protect patient data, and make the technology accessible to users with visual, cognitive, or motor impairments. Tracking patient engagement metrics, such as bounce rates and satisfaction scores, can help evaluate the return on investment.

When evaluating AR and VR for healthcare marketing, providers should weigh both the upfront costs and long-term benefits. Costs may include specialized hardware, software licensing, content development, and staff training. On the benefit side, immersive tools can shorten patient decision-making time, increase treatment acceptance rates, and reduce the need for repeat consultations by improving comprehension. For example, a clinic that invests in a $15,000 VR surgical simulation may find it pays for itself within a year if higher patient confidence leads to even a small increase in procedure bookings. However, adoption challenges include ensuring staff are comfortable using the technology, integrating it into existing workflows, and overcoming patient reluctance if the tools feel intimidating or time-consuming. Measuring ROI requires tracking engagement metrics, such as session duration, follow-up appointment rates, and patient satisfaction scores, then comparing them to baseline marketing results.

Immersive technology will not replace medical consultations, but it can make complex topics easier to understand and improve the patient journey by delivering clear, interactive, and emotionally engaging education.

Use the checklist below to plan and review each immersive asset before launch.

---

[1] https://www.linkedin.com/pulse/how-augmented-realityar-virtual-realityvr-healthcare-sector-mutabazi/. Accessed 3 Aug 2025.

Callout Box: Checklist for Immersive Content in Healthcare Marketing
Use this checklist to guide the creation and deployment of VR/AR content in your marketing strategy:

- Select content that informs and reassures patients (e.g., surgical walk-throughs, procedural guides).
- Confirm clinical accuracy by involving subject-matter experts in scripting and review.
- Feature the content prominently on your website, patient portal, and social media.
- Include a clear privacy notice and obtain consent when collecting PHI through VR/AR tools.
- Ensure accessibility for different patient populations (e.g., subtitles, simple controls).
- Measure engagement and marketing impact through metrics like session duration, conversion rate, and positive reviews.
- Monitor patient feedback or behavioral change post-exposure to assess educational effectiveness.

Map each immersive tool to a clear step in the patient journey so patients know what to do next.

Enhanced Patient Journeys Digital marketing strategies will increasingly focus on optimizing the entire patient journey, from awareness and consideration to decision and advocacy, ensuring a seamless experience across all digital touchpoints.[2] This approach involves a meticulous focus on each stage of the patient's interaction with healthcare services, from initial awareness to becoming a vocal advocate for the service or provider.

Awareness The first step involves making potential patients aware of healthcare services through targeted digital marketing campaigns. For example, a healthcare provider might use search engine optimization (SEO) techniques to appear at the top of search results for relevant health queries [9]. According to industry data, local and organic search accounts for up to 76% of web traffic in healthcare.[3] "Near-me" health searches have doubled since 2015, reflecting patient intent [10].

[2] https://webmdignite.com/blog/evolving-role-digital-marketing-healthcare#:~:text=Healthcare%20marketing%20strategies%20have%20evolved,once%20the%20ads%20were%20published. Accessed 22 Mar 2024.

[3] https://www.ruleranalytics.com/resources/healthcare-conversion-benchmark-report/. Accessed 5 Aug 2025.

Example: A regional medical provider achieved a 70% year-over-year increase in appointments from organic search after a focused local SEO program.[4]

Reported average conversion rates from SEO-driven visitors to booked appointments range from 2% to over 6%, depending on how well booking tools and clear calls to action are implemented.

Consideration Once aware, potential patients move into the consideration phase, where they evaluate the available healthcare options. Here, content marketing plays a crucial role, offering detailed information about services, patient testimonials, and expert articles to help individuals make informed decisions.[5]

Example: Search-led content programs tied to medical services have produced triple-digit gains in booked appointments. In one agency case study, a healthcare client reported a 217% year-over-year increase in booked appointments from organic search.[6]

Decision In this stage, the patient decides to choose a particular healthcare provider. Digital tools like online appointment booking systems and virtual consultations can streamline this process, making it easier for patients to act.[7]

Example: IU Health reported a 114% increase in patient utilization within 1 year across 52 departments after implementing guided online scheduling.[8]

Retention After the decision, the focus shifts to retaining the patient through excellent service delivery and engagement strategies. Instead of generic tips, practices can send flu vaccine reminders during peak season or targeted updates for chronic condition management, which have shown higher open rates and appointment bookings in physician practices.

Example: Randomized trials show that text reminders increase influenza vaccination rates compared with usual outreach, which supports targeted seasonal campaigns for established patients [11].

Advocacy The final stage is turning satisfied patients into advocates for the healthcare provider. Social media platforms serve as powerful tools for patients to share their positive experiences and recommend services to others, thereby closing the loop of the patient journey and starting a new cycle of awareness for others.[9]

---

[4] https://momenticmarketing.com/case-studies/70-yoy-increase-in-appointments-scheduled-from-organic-search?utm_source=chatgpt.com. Accessed 9 Aug 2025.

[5] https://www.linkedin.com/pulse/future-patient-acquisition-digital-marketing-online-platforms-a241e/?trk=public_post. Accessed 22 Mar 2024.

[6] https://www.techmagnate.com/healthcare-seo-case-study.html?utm_source=chatgpt.com. Accessed 9 Aug 2025.

[7] https://www.iodigital.com/en/history/raak/customer-journey-from-brand-awareness-to-brand-advocacy-in-5-steps. Accessed 22 Mar 2024.

[8] https://www.experian.com/blogs/healthcare/case-study-how-iu-health-used-guided-scheduling-to-transform-operations/?utm_source=chatgpt.com. Accessed 9 Aug 2025.

[9] https://www.growthloop.com/university/article/customer-journey-stages#:~:text=There%20are%20five%20stages%20to,fall%20in%20the%20same%20order. Accessed 22 Mar 2024.

Example: Surveys of health organizations and consumers show measurable influence from social media on provider choice. In one study, 12.5% of organizations reported attracting new patients through social media, and most consumers said a hospital's social presence affected their selection [12].

By focusing on each of these stages with tailored digital marketing strategies, healthcare providers can ensure a seamless and satisfying journey for their patients, ultimately leading to improved health outcomes and stronger patient-provider relationships.

Staying Ahead of the Curve: Continuous Learning and Adaptation

Embracing Lifelong Learning

Healthcare professionals and marketers must commit to ongoing education to keep pace with rapid technological advancements and changing patient expectations.[10,11] Continuous education and adaptation to new technologies and methodologies are critical in staying relevant and effective in one's practice.

Healthcare professionals, including doctors, nurses, and administrators, are increasingly recognizing the importance of this continuous learning journey. For instance, Mlambo (2021) highlights a competency-based approach to nurses' continuing education, emphasizing the development of clinical reasoning and leadership through reflective practice [13]. This approach underscores the necessity of integrating new knowledge and skills into daily practice to enhance patient care and outcomes.

Moreover, the digital transformation in healthcare underscores the need for marketers within the sector to stay abreast of the latest digital marketing strategies and tools. The integration of artificial intelligence, virtual reality, and personalized medicine is just a few areas where ongoing education can significantly impact healthcare delivery and marketing [14].

Embracing a growth mindset and a commitment to continuous learning enables healthcare professionals and marketers to adapt to the industry's dynamic nature. This not only benefits their career development but also, more importantly, improves patient care and engagement. As the healthcare landscape continues to evolve, the value of lifelong learning remains a constant, pivotal factor in achieving success and driving innovation in the field.

Physicians seeking to strengthen their digital marketing knowledge now have access to structured online learning programs tailored to healthcare

[10] https://www.ncbi.nlm.nih.gov/books/NBK535661/. Accessed 22 Mar 2024.

[11] https://www.precheck.com/blog/how-embrace-lifelong-learning-adapt-changing-healthcare-landscape. Accessed 22 Mar 2024.

professionals. Platforms such as the AMA Ed Hub, Coursera's Digital Marketing Specialization (offered by the University of Illinois), and Mayo Clinic's online CME courses provide flexible, accredited options that blend clinical and communication training. To overcome time constraints, many of these programs offer microlearning formats, short, focused modules that fit into busy clinical schedules. Some healthcare systems also incorporate team-based training, allowing marketing and clinical staff to learn together and align messaging strategies more effectively.

Training matters when it changes behavior and results. This section links specific upskilling efforts to measurable marketing outcomes. You will pick one tool, run a short practice on a live campaign, set a single metric, and review progress in a weekly huddle. Each example lists the skill, the result, and a simple routine to keep the gain. Use these as templates for your team.

Continuous Learning with Practical Outcomes

Training pays off when skills change daily work and numbers move.

- **GA4 attribution**

  What it is: In Google Analytics 4, attribution assigns credit for an appointment or lead to the marketing touchpoints that led to it.

  Why it matters: Smarter budget moves follow clear credit. You stop funding low-impact channels.

  How to use it:

  - Mark your primary conversion events in Admin, for example "appointment_booked."
  - Add UTM tags to every ad, email, and social link.
  - Link Google Ads to GA4.
  - Open Model comparison, review data driven vs last click, then move budget toward touchpoints that keep credit across models.

What success looks like: lower patient acquisition cost and a shorter path from first visit to booking.

CRM segmentation

What it is: In a customer relationship management system, you group patients or prospects by shared traits or behavior for targeted outreach.

Why it matters: Messages match needs, which lifts response and bookings.

How to use it:

- Create three starter segments, for example, adults with asthma, patients overdue for annual exams, and new movers within ten miles.
- Write one focused message per segment.
- Schedule automated reminders.

- Track opens, clicks, and booked appointments by segment, then keep only the winners.

What success looks like: higher open rates for targeted groups and more completed appointments from those groups.

- **Email testing**

  What it is: A simple A/B test where you change one element and see which version wins.

  Why it matters: Small improvements stack up and lift bookings.

  How to use it:

  - Pick one variable, for example, subject line.
  - Split the list, send both versions, wait 24–72 h.
  - Keep the winner and record the result in a shared playbook.
  - Test a new element next send, such as sender name or send time.

What success looks like: steady gains in open rate, click through, and booked visits month over month.

- **Call tracking and scripting**

  What it is: Call tracking assigns unique phone numbers to campaigns and records outcomes. A short script standardizes intake.

  Why it matters: You connect spend to booked visits and reduce drop-offs on the phone.

  How to use it:

  - Assign tracking numbers to major ads and landing pages.
  - Train staff on a two-step intake script, confirm reason for visit and insurance, then offer the next available slot.
  - Tag each call as booked or not booked.
  - Review five recordings each week and coach quickly.

What success looks like: higher booking to call ratio and fewer no shows.

- **Website personalization**

  What it is: The site shows content that matches visitor intent or location.

  Why it matters: Visitors see the right service and move to schedule faster.

  How to use it:

  - Identify top entry queries and the nearest clinic for each visitor.
  - Match the headline and call to action to the query.
  - Place insurance information next to the call to action.
  - Test one change at a time and watch bounce rate, time on page, forms, and calls.

What success looks like: lower bounce rate and higher form completion or call clicks on priority pages.

- **Google Business Profile operations**

  What it is: Your listing shown in Google Maps and local results, including photos, hours, reviews, and a call button.

  Why it matters: It drives calls and directions from nearby patients.

  How to use it:
  - Assign one owner for weekly updates.
  - Keep hours, services, and holiday closures current.
  - Upload fresh photos each month.
  - Request new reviews every week with a short SMS link.
  - Answer Q&A and track calls from the profile.

What success looks like: more calls from Maps and stronger visibility in the local 3-pack for target terms.

With skills in place, turn to privacy and tracking rules to keep every tactic aligned with HIPAA and recent enforcement.

Adapting to Regulatory Developments

Healthcare marketers must track frequent changes in privacy regulations that affect digital campaigns. The Health Insurance Portability and Accountability Act (HIPAA) defines strict rules for how covered entities and business associates may use patient information in marketing. HIPAA applies to any activity that encourages the use of a product or service and involves identifiable health information [15].

Using patient data for marketing requires prior written authorization. This rule includes emails, digital ads, and any outreach that refers to specific diagnoses, treatments, or providers. Even tools that appear to collect anonymous information, such as cookies, pixels, or IP addresses, may violate HIPAA if they connect to patient status or appointments [16].

In 2023 and 2024, regulators pursued enforcement actions for improper use of online tracking technologies. For example, Advocate Aurora Health agreed to pay $12.25 million to settle claims that it shared patient information with Google and Meta through tracking pixels on its patient portal [17]. The FTC and OCR also issued joint warning letters to over 130 hospitals and telehealth companies for using third-party trackers without proper safeguards [18].

These developments show the need for routine audits of marketing technologies. Healthcare organizations should examine every embedded tool, assess vendor compliance, and consult legal counsel before launching campaigns that track user behavior. Teams should not assume that data is anonymous based on platform settings alone. With the guardrails set, you can use data to target messages and improve results.

Leveraging Data for Insights

Healthcare marketers can improve campaign effectiveness by using data analytics to track patient behavior, segment audiences, and personalize messaging. Email open rates, click-through metrics, and form conversions provide valuable insights into what resonates with different patient groups.

Recent studies have shown that data-driven marketing strategies can improve patient engagement and conversion rates. For instance, campaigns that incorporate patient-specific data, such as care history or service preferences, achieve 15–25% higher open rates and stronger follow-through on appointment scheduling [19].

However, the use of healthcare data comes with ethical obligations. Organizations must clarify who owns the data collected through marketing platforms and whether patients understand how their information will be used. Transparency and informed consent are essential, especially when data originates from patient portals or is shared with external vendors. Guardrails should include de-identification where possible, formal data governance policies, and legal review of vendor contracts.

Below is a sample bar chart that illustrates open rate improvements for personalized versus generic email campaigns in healthcare. This kind of visual feedback can help marketing teams refine their strategies based on measurable outcomes.

Now define the numbers you will track to judge progress.

Measuring Success: KPIs and Analytics in Healthcare Marketing

To get value from healthcare marketing, you need to measure what matters. Without benchmarks, even the best campaigns waste time and money.

Key Performance Indicators (KPIs)

KPIs help you decide what's working and where to adjust. The most useful ones vary by goal and channel, but in clinical settings, consider the following:

- **Appointment Conversion Rate**: What percentage of people who click an ad or visit your site schedule an appointment?
- **Patient Acquisition Cost (PAC)**: How much do you spend to bring in each new patient?
- **Email Metrics**: Open rates and click-through rates show how well your subject lines and messages perform.
- **Bounce Rate**: If people leave your website quickly, your message or layout may be off.

- **Call Tracking Volume**: Track how many calls come from specific ads or landing pages.
- **Local SEO Performance**: Are you showing up in Google's "local 3-pack?" The local 3-pack refers to the top three business listings that appear in a Google Map search result, which receive the highest visibility and click-through rates for local searches. They appear in searches like "dermatologist near me"?

For a practical tool to begin tracking these metrics, see Appendix: Sample KPI Tracking Template. To make sense of these KPIs and translate numbers into decisions, healthcare marketers need reliable, user-friendly analytics tools.

Data Interpretation: Turning KPIs into Decisions

Numbers matter only if they change what you do. Use this quick loop.

1. Confirm the signal. Compare with last month and with the same month last year. Check sample size and tracking integrity. No high fives until you know why a metric moved.
2. Find the driver. Break results by channel, campaign, keyword, device, location, and new versus returning visitors.
3. Choose one action. If appointment conversion drops, review the booking flow. Use fewer fields, clarify insurance language, and add a one-click call option.
4. Test fast. Define hypothesis, metric, target, and time window. Run an A/B test with equal traffic.
5. Close the loop. Log the change, revisit the metric in 1 to 2 weeks, then scale or roll back.

Metric to Action Cheat Sheet

Before you act, confirm the signal. Compare with last month and the same month last year. Check the sample size and that tracking works.

- **High bounce rate**

Likely issues: slow load, weak match between search query and page, and mobile layout problems.

Actions: run a speed test, compress images, and remove pop-ups. Match the headline to the query. Move the call to action above the fold. Check mobile readability.

- **Strong traffic with low appointment conversion**

Likely issues: friction in scheduling, unclear insurance or pricing, weak trust signals, broken thank you tracking.

Actions: shorten the form, show insurance accepted, add a phone option, add reviews and clinician bios, and fix the conversion tag on the confirmation page.

- **Rising patient acquisition cost (PAC)**

Likely issues: poor targeting, low ad relevance, weak landing pages, and budget spread across too many campaigns.

Actions: tighten geography and time windows. Pause low performers. Match ad copy to the keyword. Align landing page copy with the ad. Shift spend to the highest converting campaigns.

- **Low email open rate**

Likely issues: weak subject line or sender name, poor send time, stale list, and deliverability problems.

Actions: A/B test the subject line and sender name. Adjust send time. Remove inactive addresses. Authenticate your domain with SPF, DKIM, and DMARC.

- **Low click-through rate (CTR)**

Likely issues: unclear offer or snippet, weak creative, or wrong audience.

Actions: refresh titles and meta descriptions. Add structured data for FAQs or reviews. Tighten audience targeting. Sharpen the call to action.

- **Drop in local 3-pack rank**

The local three pack is the three Google Map listings shown for a local search.

Likely issues: few new reviews, incomplete profile, inconsistent name, address, and phone across directories, and weak local content.

Actions: request new Google reviews each week. Update Google Business Profile details, services, and photos. Fix NAP consistency. Create location pages with driving directions and parking information.

With KPIs set, choose tools that capture these measures without extra steps for your team.

Using Analytics Tools

You don't need advanced training to track results. These tools help physicians make sense of the data:

- **Google Analytics** shows who visits your site, how they got there, and what they do.
- **CallRail** tracks inbound calls, duration, and whether they lead to appointments.
- **CRM Dashboards with EHR Integration** (e.g., Salesforce Health Cloud) combine marketing and clinical data to spot trends and gaps.

These tools become even more valuable when combined with actionable insights drawn from real practice experiences. The example below shows how these tools drive measurable results.

Figure 13.1 summarizes core analytics tools and what you learn from each.

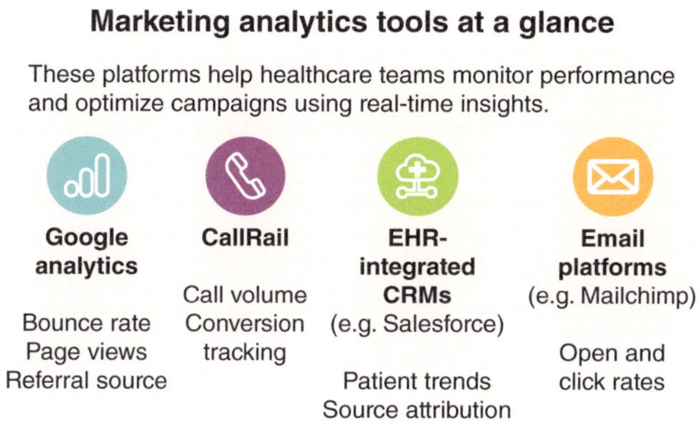

Fig. 13.1 Essential marketing analytics tools and their key functions to help healthcare providers track campaign performance and improve patient engagement. (Image by Quinn/ChatGPT)

Real-World Example

A dermatology clinic in Chicago partnered with a digital agency to optimize its Google Ads strategy. Within 3 months, patient visits increased by 17% and new patient intake rose by 16%, largely due to improved targeting and clearer calls to action [20]. In another case, a multi-location dermatology group in Oregon enhanced both SEO and PPC campaigns. This resulted in a 119% increase in organic website traffic, a 69% jump in organic clicks, and a 25% reduction in cost per patient acquisition [21].

This example highlights how data-driven adjustments, even on a small scale, can yield measurable improvements in patient acquisition and engagement. These results show why routine measurement matters.

Takeaway

Track your data regularly. It helps you move quickly, test smarter, and spend wisely. When you measure outcomes, you build better campaigns and more confident decisions, without guessing. The conclusion brings the chapter together and lists the actions to take next.

Conclusion

Success in healthcare marketing depends on continuous tracking, analysis, and adaptation. By selecting the right KPIs and applying insights from accessible analytics tools, healthcare providers can make smarter decisions, reduce waste, and increase patient engagement. Data-backed strategies do more than improve marketing performance; they help align outreach efforts with patient needs, organizational goals, and ethical standards. In a competitive digital environment, the ability to measure and respond quickly is not optional. It's essential.

Appendix: Sample KPI Tracking Template

About This Template

This KPI Tracking Template helps healthcare practices monitor and evaluate the performance of their marketing campaigns over time. Each row identifies a key performance indicator (KPI) that reflects specific marketing goals, such as increasing patient appointments, reducing acquisition costs, or improving engagement rates. The columns allow teams to track target and actual values, note changes month to month, and document recommended actions. By reviewing this data regularly, practices can make informed adjustments that improve both marketing efficiency and patient outreach.

KPI	Definition	Target value	Current value	Change from last month	Notes/action steps
Appointment conversion rate	% of leads who schedule appointments after engaging with marketing	35%	28%	−3%	Improve follow-up messaging
Patient acquisition cost (PAC)	Marketing spend ÷ new patients acquired	$100	$125	+$10	Optimize ad targeting
Email open rate	% of recipients who open marketing emails	30%	32%	+2%	Subject line performing well
Click-through rate (CTR)	% of email or ad viewers who click a link	5%	4.2%	−0.8%	Test new CTA language
Bounce rate	% of website visitors who leave without taking action	<50%	61%	+5%	Review landing page content
Call volume (tracked)	Number of inbound calls from marketing efforts	200/month	175	−12.5%	Refresh ad creative
Local SEO rank	Position in Google local 3-pack for target keywords	Top 3	Position 4	−1	Encourage more Google reviews

References

1. Deloitte. 2025 global health care sector outlook: the future of health. Deloitte Insights; 2025. https://www.deloitte.com/us/en/insights/industry/health-care/life-sciences-and-health-care-industry-outlooks/2025-global-health-care-executive-outlook.html. Accessed 3 Aug 2025.
2. Centers for Medicare & Medicaid Services. Telehealth: delivering care safely during covid-19 and beyond; 2023. https://www.cms.gov/newsroom/fact-sheets/medicare-telemedicine-health-care-provider-fact-sheet. Accessed 4 Aug 2025
3. Santosa KB, Cederna PS. Commentary on: trends and challenges of telehealth in an academic institution: the unforeseen benefits of the COVID-19 global pandemic. Aesthet Surg J. 2021;41(1):119–21. https://doi.org/10.1093/asj/sjaa272.
4. Boston Consulting Group. The future of digital health: a roadmap for 2025 and beyond; 2024. https://www.bcg.com/publications/2024/the-future-of-digital-health. Accessed 4 Aug 2025
5. Johnson KB, Wei WQ, Weeraratne D, Frisse ME, Misulis K, Rhee K, Zhao J, Snowdon JL. Precision medicine, AI, and the future of personalized health care. Clin Transl Sci. 2021;14(1):86–93. https://doi.org/10.1111/cts.12884.
6. Iqbal AI, Aamir A, Hammad A, Hafsa H, Basit A, Oduoye MO, Anis MW, Ahmed S, Younus MI, Jabeen S. Immersive technologies in healthcare: an in-depth exploration of virtual reality and augmented reality in enhancing patient care, medical education, and training paradigms. J Prim Care Community Health. 2024;15:21501319241293311. https://doi.org/10.1177/21501319241293311.
7. Tang YM, Chau KY, Kwok A, Zhu T, Ma X. A systematic review of immersive technology applications for medical practice and education: trends, application areas, recipients, teaching contents, evaluation methods, and performance. Educ Res Rev. 2022;35:100429. https://doi.org/10.1016/j.edurev.2021.100429.
8. Lewis KO, Popov V, Fatima SS. From static web to metaverse: reinventing medical education in the post-pandemic era. Ann Med. 2024;56(1):2305694. https://doi.org/10.1080/07853890.2024.2305694.
9. Purcarea VL. The impact of marketing strategies in healthcare systems. J Med Life. 2019;12(2):93–6. https://doi.org/10.25122/jml-2019-1003.
10. Think With Google. https://business.google.com/us/think/ (n.d.)
11. Patel MS, Milkman KL, Gandhi L, Graci HN, Gromet D, Ho H, Kay JS, Lee TW, Rothschild J, Akinola M, Beshears J, Bogard JE, Buttenheim A, Chabris C, Chapman GB, Choi JJ, Dai H, Fox CR, Goren A, Hilchey MD, Hmurovic J, John LK, Karlan D, Kim M, Laibson D, Lamberton C, Madrian BC, Meyer MN, Modanu M, Nam J, Rogers T, Rondina R, Saccardo S, Shermohammed M, Soman D, Sparks J, Warren C, Weber M, Berman R, Evans CN, Lee SH, Snider CK, Tsukayama E, Van den Bulte C, Volpp KG, Duckworth AL. A randomized trial of Behavioral nudges delivered through text messages to increase influenza vaccination among patients with an upcoming primary care visit. Am J Health Promot. 2023;37(3):324–32. https://doi.org/10.1177/08901171221131021.
12. Ventola CL. Social media and health care professionals: benefits, risks, and best practices. Pharmacy Ther. 2014;39(7):491.
13. Mlambo M, Silén C, McGrath C. Lifelong learning and nurses' continuing professional development, a metasynthesis of the literature. BMC Nurs. 2021;20(1):62. https://doi.org/10.1186/s12912-021-00579-2.
14. Kuwabara A, Su S, Krauss J. Utilizing digital health technologies for patient education in lifestyle medicine. Am J Lifestyle Med. 2019;14(2):137–42. https://doi.org/10.1177/1559827619892547.
15. U.S. Department of Health and Human Services. HIPAA privacy rule and marketing; 2023. https://www.hhs.gov/hipaa/for-professionals/privacy/guidance/marketing/index.html
16. Office for Civil Rights. Use of online tracking technologies by HIPAA covered entities and business associates; 2023, December 1. https://www.hhs.gov/hipaa/for-professionals/privacy/guidance/hipaa-online-tracking/index.html

17. U.S. District Court, Northern District of Illinois. Class action settlement in In re Advocate Aurora Health Pixel Litigation, Case No. 1:22-cv-07156; 2023. Retrieved from https://www.advocateauroraclassaction.com/
18. Federal Trade Commission & Office for Civil Rights. FTC and HHS Office for Civil Rights issue warning letters about privacy violations from online tracking technologies; 2023, July 20. https://www.ftc.gov/news-events/news/press-releases/2023/07/ftc-hhs-office-civil-rights-issue-warning-letters-about-privacy-violations-online-tracking-technologies
19. Salesforce. Trends in healthcare marketing: Patient engagement in a digital-first era; 2024. https://www.salesforce.com/resources/articles/healthcare-marketing-trends-2024/
20. MVotion. Growing a dermatology practice with digital ads: case study; 2021. Retrieved from https://www.mvotion.com/growingadermatologypracticewithdigitalads. Accessed 4 Aug 2025.
21. Cardinal Digital Marketing. SEO and PPC case study: Center for Dermatology & Laser Surgery; n.d.. Retrieved from https://www.cardinaldigitalmarketing.com/healthcare-case-study/center-for-dermatology-laser-surgery/. Accessed 4 Aug 2025.

Chapter 14
Innovative Content Marketing Strategies for Healthcare

This chapter explores how healthcare organizations can use innovative content formats to engage patients, build trust, and strengthen their digital presence. You will learn how to apply tools like infographics, videos, podcasts, generative AI, chatbots, and virtual assistants, along with strategies to ensure your content is accurate, accessible, and measurable. The chapter also provides best practices, sample prompts, and performance metrics to help you plan and evaluate effective campaigns.

Attracting the attention of patients on digital platforms can be challenging. Content marketing has emerged as a powerful tool to engage audiences, disseminate information, and ultimately improve health outcomes. By leveraging various types of content such as videos, infographics, and podcasts, healthcare providers can connect with their audience in a more personalized and impactful way.

The following content formats represent the most commonly used tools in healthcare marketing today. Tie each format to a measurable outcome, for example, booked appointments or call volume, not awareness alone. Define success before you start. Pick one format, set one metric, and review results weekly. Start with a video to explain care fast and build trust online.

Videos

Videos have become an indispensable tool in healthcare digital marketing, offering a dynamic way to convey complex medical information in an easily digestible format. Healthcare marketing videos are ideal for posting on your social media platforms and your website's homepage. They offer a distinctive way to guide patients through your site and ensure their initial visit is engaging and informative. There are several innovative ways videos can be integrated into content marketing strategies to engage and educate patients effectively.

C. Quinn, *The Digital Practice*, https://doi.org/10.1007/978-3-032-07745-5_14

1. *Patient Testimonials:* Sharing real-life stories of patients who have undergone treatments or procedures can build trust and empathy among viewers. These testimonials can highlight the quality of care provided by healthcare facilities and the positive impact on patients' lives.[1]
2. *Provider Profiles:* Videos that introduce healthcare providers, showcasing their expertise, personality, and approach to patient care, can help build a connection with potential patients before they even step into the office. This personal touch can be crucial in choosing a healthcare provider.[2]
3. *Educational Series:* A series of short videos covering various health topics, from disease prevention to wellness tips, can position a healthcare brand as a thought leader in the industry. For instance, explainer videos on health conditions, procedures, and preventive care can significantly enhance patient understanding and engagement. These videos can be shared across social media platforms, websites, and email newsletters to reach a wider audience.[3]
4. *Behind-the-Scenes Tours:* Offering a glimpse into the day-to-day operations of healthcare facilities, including the technology and care processes, can demystify the patient experience and alleviate anxieties about medical procedures.[4]

Video content holds unique value because it combines visual clarity with emotional resonance, helping patients understand complex topics and feel more connected to their care. For each video posted, track one measurable action such as Schedule Now clicks, Call button taps, or form starts. Place a "Schedule An Appointment" button directly below the video player on the page, repeat the same button at the top and end of the transcript, and add a Call button on mobile that dials the clinic.

Follow with infographics to compress key facts into quick, shareable visuals.

Infographics

An infographic visually presents data, information, or knowledge using illustrations, icons, charts, and minimal text. Healthcare marketers use infographics to explain topics, summarize statistics, or highlight patterns in a format that is quick to interpret and visually engaging.

Visuals help patients interpret complex information more quickly than text alone [1]. Because of this, healthcare professionals often use infographics to

[1] https://vidico.com/news/8-examples-of-well-done-healthcare-marketing-videos/. Accessed 24 Mar 2024.

[2] https://ahamediagroup.com/blog/healthcare-video-marketing-strategies-examples/. Accessed 24 Mar 2024.

[3] https://comradeweb.com/blog/healthcare-video-marketing/. Accessed 24 Mar 2024.

[4] https://healthcare.report/articles/healthcare-video-marketing-strategies-and-best-practices. Accessed 24 Mar 2024.

communicate messages like disease prevention steps or lifestyle benefits, especially when promoting health literacy [2]. These tools support storytelling and data presentation by combining words and images in ways that highlight essential insights.

For example, one infographic might walk patients through how vaccination works and why it matters. Another could illustrate how diet, exercise, and smoking affect chronic diseases like heart disease and diabetes. These examples show how visual content improves patient understanding and supports proactive health decisions. Infographics are particularly valuable when working with audiences who have varying levels of health literacy, helping healthcare providers deliver essential information clearly and efficiently. Add alt text for screen readers, use high contrast colors that meet WCAG 2.1 AA, and label axes and data points in plain language. Alt text means a short written description attached to an image. Screen readers read it aloud. Browsers show it if the image fails to load. It helps patients who use assistive technology, and it improves search indexing.

Use podcasts when you need minimal depth and a steady clinical voice between visits.

Podcasts

Podcasts have emerged as a powerful tool to engage audiences on a more personal and in-depth level. This medium provides an unparalleled opportunity to explore health topics extensively, featuring discussions with medical experts, sharing patient success stories, and disseminating the latest research findings. The intimate nature of podcasts, where listeners often feel a direct connection with the hosts and guests, can significantly enhance the sense of community and support among individuals facing similar health challenges [3].

Podcasts are relatively low-cost to produce and can be repurposed into blog posts, infographics, and social media clips. Starting a podcast does not require a professional studio. Entry-level microphones like the Blue Yeti and free editing tools such as Audacity offer a good starting point. For distribution, platforms like Spotify, Apple Podcasts, and Podbean make publishing accessible and widely reachable.

Podcasts in healthcare content marketing can serve multiple purposes. For instance, a series might focus on preventive healthcare, offering listeners practical advice on maintaining a healthy lifestyle from renowned nutritionists and fitness experts. For example, Mayo Clinic produces “The Mayo Clinic Q&A,” a podcast that features physicians and researchers discussing timely health topics. Podcasts enhance patient connection by offering consistent access to provider voices, reinforcing transparency and availability.

Another podcast could delve into chronic disease management, providing comfort and actionable guidance to those navigating the complexities of long-term health conditions. Additionally, podcasts that feature patient stories of recovery and resilience can be incredibly inspiring, offering hope and motivation to others on their health journey.

By integrating podcasts into their content marketing strategies, healthcare organizations can educate and inform their audience. This approach enables deeper engagement with content that is both informative and emotionally resonant, making it a vital component of modern healthcare marketing strategies.

Sidebar: Podcast Quick Start: Cost and Two-Hour Setup

Typical cost

- USB microphone, 80–200 dollars
- Desktop boom arm and pop filter, 30–60 dollars
- Closed-back headphones, 50–120 dollars
- Basic acoustic panel or portable shield, 50–100 dollars
- Recording and editing software, free with Audacity or GarageBand, or 15–30 dollars per month for tools like Descript
- Podcast hosting, 0–20 dollars per month

Two-hour setup path

- 0:00–0:10 outline three points and the call to action
- 0:10–0:20 connect the mic, set input level so peaks sit near minus 12 dB
- 0:20–0:45 record intro and main segment, aim for 12–15 min total
- 0:45–1:15 edit, remove long pauses, apply noise reduction, normalize to about minus 16 LUFS
- 1:15–1:25 export MP3, 96 to 128 kbps mono, add title and show notes with a Schedule Appointment link
- 1:25–1:40 upload to a podcast host, confirm RSS feed
- 1:40–2:00 embed the episode on your website, add a short transcript, post a 30-second teaser on social

HIPAA reminder

Keep stories de-identified. Get written consent for any patient voice or detail.

Success check

Track downloads per episode, completion rate, clicks to scheduling, and calls from the episode page.

To keep production on schedule, use generative AI for first drafts and outlines, then review for accuracy.

Using Generative AI to Streamline Content Creation

Generative AI tools such as ChatGPT[5], Jasper[6], Claude[7], and DocsGPT [4][8] allow physicians and clinic staff to produce content quickly. You can prompt these tools to generate outlines, blog posts, email headlines, social media captions, or even short video scripts. Some platforms now include medical-specific options designed for professional use, such as DocsGPT for doctors or Hippocratic AI for clinical applications.[9]

These tools speed up production but do not replace clinical expertise. Generative AI models work by predicting likely word patterns, not by understanding medical science. They often include errors, omit important context, or present outdated or incorrect information. Use them as a starting point, not a final draft.

How to Use Generative AI Safely and Effectively

- **Write clear, specific prompts.** Example: "Create a 300-word blog post explaining the difference between Type 1 and Type 2 diabetes for patients." You can also ask the tool to summarize or rewrite text you already have.
- **Always review and edit the output.** Check for clinical accuracy, update any outdated references, and make sure the tone reflects your practice's style.
- **Do not input patient information.** Most free tools are not HIPAA-compliant and store prompts on external servers. Use de-identified data or a secure tool with privacy assurances.
- **Stick to your publishing schedule.** Use AI to draft content ahead of time, then revise and approve before posting. To streamline planning and maintain consistency, see Appendix for a sample content calendar template that helps organize digital campaigns across multiple platforms.
- **Include a disclaimer when needed.** Let patients know if a blog post or newsletter was created with AI assistance.

Small practices benefit the most. These tools can reduce the time needed to write content while still letting your clinical voice come through. Used wisely, AI supports consistency, efficiency, and scale without compromising quality or compliance. During review, check for model drift and stale facts by verifying dates on statistics and clinical guidance, and record the model name and version used. Use

[5] https://chatgpt.com/. Accessed 5 Aug 2025.

[6] https://www.jasper.ai/. Accessed 5 Aug 2025.

[7] https://claude.ai/. Accessed 5 Aug 2025.

[8] https://www.docsgpt.cloud/. Accessed 5 Aug 2025.

[9] https://www.hippocraticai.com/?utm_source=chatgpt.com. Accessed 5 Aug 2025.

the content calendar in Appendix to set deadlines, assign owners, and keep a steady weekly cadence.

Next, apply AI to video with avatars that deliver your script without cameras or actors.

Video content is evolving quickly, and AI avatar tools now allow clinics to create professional videos without cameras or actors.

Sidebar: Sample AI Prompts for Busy Healthcare Teams

You don't need technical skills to get useful content from generative AI. Start by giving the tool clear, specific instructions. Here are examples physicians and clinic staff can use to generate first drafts quickly:

For Blog Posts

- Write a 300-word blog post explaining what an A1c test measures and why it matters.
- Draft a short article on seasonal allergies and treatment options for adults.

For Social Media

- Create five Instagram captions to promote flu shots at a family medicine clinic.
- Write a tweet reminding patients to schedule their annual physical.

For Patient Education

- Summarize the risks and benefits of colon cancer screening in plain language.
- Create a one-paragraph explanation of prediabetes for a handout.

For Internal Use

- Draft an email to patients announcing extended clinic hours on Saturdays.
- Write a reminder script for front-desk staff to use when patients cancel too often.

Editing Help

- Rewrite this paragraph to sound warmer and more conversational.
- Summarize this medical journal abstract for a patient newsletter.

Tip: Always read the results closely and revise as needed. Think of AI as a fast-writing intern; it works quickly but still needs supervision.

Video Marketing and AI Avatars

You no longer need a camera crew or studio to produce professional-looking video. AI avatar tools like **Synthesia** and **HeyGen** enable you to create digital presenters that speak your script aloud. These tools convert typed text into videos featuring lifelike voices and facial expressions.

Physicians use these videos for:

- Patient education clips shown in waiting rooms
- Short explainer videos shared on social media
- Website content introducing services or procedures
- Multilingual instructions without needing human translation

To get started, choose a tool, select an avatar, and paste your script into the platform. The software generates a video in minutes. You can then download and share it wherever your audience will see it, on your website, in emails, or on social channels.

For example, a pediatrician might create a 60-s video explaining the importance of the flu vaccine, recorded in English and Spanish, and post it on the practice's Facebook page.

Be cautious about what information you include. These tools are not HIPAA-compliant, so do not use real patient names, cases, or any protected data. Stick to general information and public health messages.

Many platforms offer free trials, but full features are often available only with a subscription. If your office doesn't have a marketing team, these tools still let you produce polished videos quickly and affordably.

Done well, AI avatars deliver clear, consistent, and multilingual messages that help patients feel informed and supported, especially in sensitive or complex topics, without requiring providers to appear on camera. See Fig. 14.1 for a summary of formats and best practices. Track outcomes for avatar videos, including completion rate for clips under 60 s, clicks to schedule after viewing, and reach by language for the same script.

After you publish, support patients with conversational tools that answer questions and route requests.

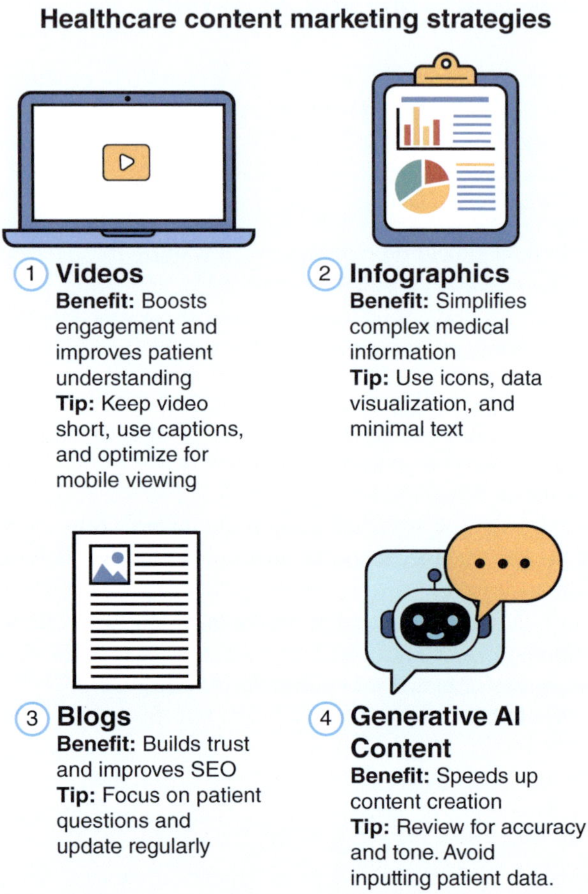

Fig. 14.1 Summary of the most effective healthcare content marketing strategies, including video, infographics, blogs, and AI-generated content, along with their key benefits and best practices. (Image Quinn/ChatGPT)

AI Chatbots and Voice Assistants in Clinical Practice

Digital tools like AI chatbots and voice assistants are becoming essential parts of modern medical practices. They improve access, streamline communication, and create a more responsive experience for patients. These technologies are increasingly used in both large systems and small clinics, but they serve different purposes and require different strategies for implementation. Use these tools only for general information and always avoid patient data. Do not collect or store protected health information in chat logs. Place a clear privacy notice above the chat window.

Use Table 14.1 to weigh benefits and drawbacks before you choose a platform.

Table 14.1 Benefits and drawbacks of chatbots and voice assistants in healthcare marketing

Feature	Benefits	Drawbacks
24/7 Availability	Answers basic questions anytime, improving access and responsiveness	May frustrate users if answers are incomplete or off-topic
Efficiency	Handles routine tasks like appointment scheduling or FAQs	Limited in handling complex or nuanced queries
Cost-Effective	Reduces administrative workload and call volume	Setup and maintenance costs may be high, especially for custom integrations
Personalization	Can tailor responses based on patient inputs	Risk of sounding robotic or impersonal if not well-designed
Data Collection	Captures valuable user data for marketing insights	Raises privacy concerns and must comply with HIPAA
Multilingual Support	Expands accessibility for diverse populations	Quality varies depending on language and dialect accuracy

AI Chatbots

What is a chatbot?

A chatbot is a software application that interacts with users through written messages. It responds to questions, provides information, and completes tasks based on user input. In a medical practice, chatbots typically reside on the practice website, within the patient portal, or in a dedicated mobile app.

How do patients access them?

Patients usually encounter chatbots when they visit your website or use your mobile app. A small pop-up or message box often invites them to ask questions or schedule services. Some systems can also integrate with Facebook Messenger or SMS platforms, offering multiple access points.

Everyday tasks handled by chatbots include:

- Appointment scheduling
- Prescription refill requests
- Clinic hours or location information
- Pre-visit symptom screening
- Answers to FAQs

Who develops and installs them?

Vendors such as **HealthTap**, **Gyant**, and **Mediktor** offer prebuilt chatbot platforms tailored for medical settings. These tools can often be branded for your practice and integrated with your EHR or scheduling system. Installation usually requires collaboration between the vendor and your website developer or IT team. Some EHR systems, like **Athenahealth** or **Epic**, now offer built-in or partner chatbot tools as well.

Examples of successful use:

A primary care network in California integrated Gyant's chatbot across its websites. Within six months, over 30% of appointment requests came through the chatbot, with high patient satisfaction scores [5]. The tool also reduced call center volume [6].

Challenges:

- **Cost:** Monthly fees range from a few hundred to several thousand dollars, depending on features and customization.
- **Integration:** Some chatbots struggle to link smoothly with older EHRs.
- **Maintenance:** Content must be updated regularly to stay accurate and relevant. Without monitoring, outdated responses can confuse patients or provide incorrect information.

Sidebar: Top Content Tools for Busy Clinics

This sidebar offers a quick-reference list of practical tools to support content creation and distribution for physicians and healthcare marketers.

AI Writing Tools

- ChatGPT
- Jasper
- Claude

Design Tools

- Canva
- Adobe Express

Video Tools

- Animoto
- InVideo

Podcast Hosting

- Buzzsprout
- Spotify for Podcasters

Infographic Tools

- Venngage
- Piktochart

Reminder: Always verify that any tool used to create or distribute patient-related content complies with HIPAA or other applicable privacy regulations.

AI Voice Assistants

What is a voice assistant?

A voice assistant is an AI system that responds to spoken commands. These tools help physicians complete tasks using their voice, reducing the need to type or click through menus. Some tools also listen during patient visits and generate documentation automatically.

Where are they used?

Voice assistants are typically used on smartphones, tablets, or desktop microphones during clinical encounters. Some also integrate directly into the EHR.

What can they do?

- Draft clinical notes
- Retrieve lab results or medication lists
- Place orders or referrals
- Schedule follow-up appointments
- File documentation after a visit

Popular platforms include:

- **Nuance DAX (Dragon Ambient eXperience)**[10]
- **Suki Assistant**
- **Amazon Clinic Voice (in pilot stages)**

Example of use:

The Permanente Medical Group implemented Nuance's Dragon Ambient eXperience (DAX) across multiple specialties. Physicians reported a 50–75% reduction in clinical documentation time. Additionally, 88% of providers indicated higher satisfaction with their documentation experience. These improvements were most notable in complex clinical scenarios, where DAX helped reduce administrative burden and streamline the creation of accurate, structured notes [7, 8].

Challenges:

- **Cost:** These tools can be expensive, especially for smaller practices. Subscription fees, hardware requirements, and training time should be considered.
- **Installation:** Some tools require setup across all exam rooms or provider devices. Integration with your EHR is key for smooth use.
- **Accuracy:** While voice assistants are improving, they still require review and editing. Accents, background noise, or clinical complexity may reduce transcription quality.

The next section ties these tools to access, satisfaction, and referrals.

[10] https://tali.ai/resources/nuance-dax-how-ai-can-transform-your-clinical-documentation?utm_source=chatgpt.com. Accessed 5 Aug 2025.

How These Tools Help Market Your Practice

Chatbots create a seamless front-door experience. They make it easy for prospective patients to schedule visits or ask questions without calling. This improves satisfaction, builds trust, and signals that your practice values convenience.

Voice assistants work in the background, helping you operate more efficiently. When physicians are less burdened by documentation, they can respond faster to messages, offer same-day visits, and spend more time with each patient. These improvements directly support your brand and can generate word-of-mouth referrals.

Both tools reflect a modern, tech-savvy image of your practice, one that patients increasingly expect. While the initial setup requires effort, the long-term benefits in terms of efficiency and patient satisfaction can be significant.

Before you choose a platform, confirm your goal, for example, after-hours coverage or intake. Verify EHR integration for identity, scheduling, and secure messaging.

Assign staffing for content review and escalation. Set one success metric such as containment rate, median response time, or booked appointments.

See Fig. 14.2 to compare chatbots and voice assistants at a glance.

With tools selected, follow these best practices to keep each message accurate, timely, and easy to read.

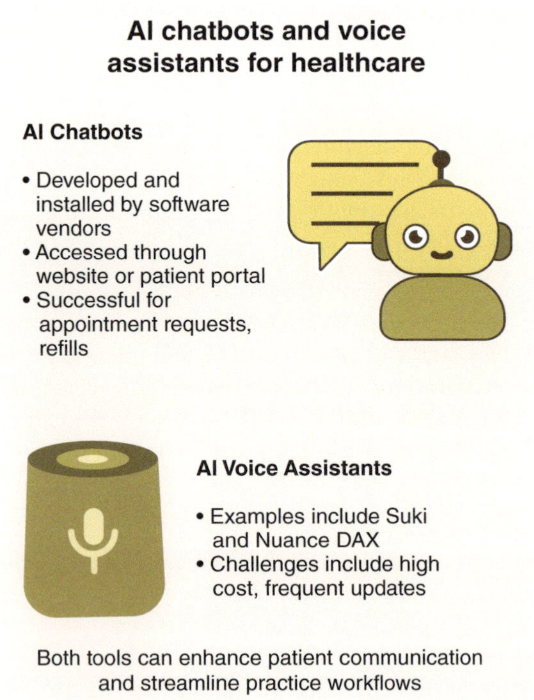

Fig. 14.2 Choosing between AI chatbots and voice assistants in your practice

Best Practices for Creating Engaging and Informative Content

Once you choose a format, make sure your content follows best practices for effectiveness.

Creating engaging and informative content is crucial for healthcare digital marketing. Here's how to implement these best practices effectively:

1. *Understand Your Audience*: Effective healthcare marketing begins with knowing your audience. Instead of trying to appeal to everyone, focus on the specific needs and concerns of your patient population. This targeted approach improves engagement and ensures your content reaches the people most likely to benefit from your services. Start by researching patient demographics, behaviors, and health interests. Use surveys, website analytics, or direct conversations on social media to identify what your audience cares about. Then tailor your content to address those concerns clearly and consistently.[11]
2. *Maintain Accuracy and Reliability*: In healthcare, the accuracy of information can have significant implications. Collaborate with other healthcare professionals to verify the content before publication. Additionally, citing reputable sources such as peer-reviewed journals or official health organization guidelines can enhance the credibility of your content.[12]
3. *Make It Current*: To attract increased audience interest, aim to link your material to prevailing health trends, significant medical news, or topics currently engaging your patient community. This approach not only makes your content more relatable but also positions your practice as a timely and informed voice in the healthcare conversation. By aligning your content with what's currently relevant to your patients, you're more likely to attract their attention and foster meaningful engagement.
4. *Make It Accessible:* Your audience is unlikely to have a medical background. Use layman's terms to explain complex medical terms and conditions. Incorporating visuals like infographics or explanatory videos can also help break down complex information into digestible pieces, making it easier for people of all education levels to understand.
5. *Encourage Interaction:* Create content that prompts your audience to interact. This could be through asking questions at the end of your articles, creating polls on social media, or encouraging comments on blog posts. Interaction not only boosts engagement but also provides valuable feedback and insights into your audience's needs and preferences.

[11] https://www.linkedin.com/advice/3/what-some-best-practices-creating-informative-educational-9jhue. Accessed 24 Mar 2024.

[12] https://www.forbes.com/sites/forbesagencycouncil/2020/02/27/15-best-practices-for-producing-quality-engaging-content/?sh=5ad786da46ed. Accessed 24 Mar 2024.

By implementing these strategies, healthcare organizations can create content that not only informs and educates but also builds a strong, engaged community around their brand.

Now track results and adjust based on the numbers.

Measuring Effectiveness

In healthcare, the effectiveness of content marketing strategies is pivotal for both patient engagement and organizational growth. To gauge this impact, healthcare providers can rely on a variety of metrics. Website traffic serves as a primary indicator, revealing the volume of visitors drawn to the healthcare provider's online resources. This, when analyzed, can offer insights into the reach and appeal of the content presented.[13] For KPI definitions and a template, see Chap. 13.

Here's how to measure it effectively:

1. **Session Counts**: Track the total number of visits to your healthcare website over specific intervals, weekly, monthly, etc. This metric helps gauge the impact of digital marketing efforts like search engine optimization (SEO) and pay-per-click (PPC) advertising.
2. **Conversion Rates**: For healthcare platforms offering services or appointments online, monitoring the percentage of visitors who take action (e.g., booking an appointment) is vital.
3. **Traffic Sources**: Identify how visitors are finding your website—be it through search engines, direct visits, social media, or referral links. This insight allows for targeted improvements in marketing channels that are underperforming.
4. **Bounce Rate**: The bounce rate, or the proportion of visitors who leave after viewing just one page, should ideally be low. A high bounce rate might indicate that the content is not meeting the needs or expectations of the visitors, necessitating content optimization.

Social Media Engagement Social media engagement is a measure of all interactions with your social media content. It serves as a crucial indicator of content marketing success, particularly in the healthcare sector. Metrics such as likes, shares, retweets, comments, and overall interaction rates offer insights into audience preferences and content performance.[14] These metrics not only reflect the content's ability to engage and captivate the audience but also its potential to foster meaningful conversations and community building around health topics.

For healthcare providers, understanding these engagement patterns can guide the creation of more relevant and impactful content. For example, a high number of

[13] https://airfocus.com/glossary/what-is-website-traffic/. Accessed 24 Mar 2024.

[14] https://socialclimb.com/blog/how-to-measure-the-performance-of-your-healthcare-marketing-strategy/. Accessed 24 Mar 2024.

shares on an infographic about preventive healthcare measures could indicate a strong interest in practical health tips, suggesting a direction for future content creation. Similarly, comments and questions on a video about a specific treatment can highlight areas where patients seek more information, guiding healthcare providers to address these topics in more depth.

Moreover, when healthcare providers actively participate in conversations, respond to comments, and share valuable information, they demonstrate their commitment to patient education and support. This not only strengthens the provider–patient relationship but also amplifies the reach of their content through organic sharing, ultimately contributing to a healthier and more informed community.

Social media engagement further illuminates the effectiveness of content marketing efforts.

Patient Feedback In healthcare digital marketing, patient feedback is defined as the collection of patients' opinions, experiences, and perceptions regarding the quality of care and services they receive from healthcare providers. This feedback can be gathered through various digital channels, including online surveys, social media platforms, review websites, and digital feedback systems. The primary goal of collecting patient feedback in digital marketing is to identify areas for improvement, enhance patient satisfaction, and ultimately, foster patient loyalty and referrals.[15,16]

Patient feedback serves as a crucial component of a healthcare provider's online reputation management strategy. It offers insights into what patients value in their care experience and highlights opportunities for healthcare organizations to adjust their services and communication strategies to better meet patient needs. For instance, surveys can reveal patient satisfaction levels and areas for improvement in service delivery or information clarity.[17] Reviews on platforms like Google Business Reviews offer public testimonials of patient experiences, influencing potential patients' choices and trust in a healthcare provider. By leveraging patient feedback, healthcare providers can make informed decisions to improve their marketing efforts, tailor their services to patient expectations, and promote a positive image that attracts new patients while retaining existing ones.[18]

[15] https://www.qualtrics.com/experience-management/industry/patient-feedback/. Accessed 24 Mar 2024.

[16] https://www.questionpro.com/blog/patient-feedback/. Accessed 24 Mar 2024.

[17] https://www.drcatalyst.com/blog/incorporating-patient-feedback-into-your-marketing-plan#:~:text=This%20is%20called%20feedback%20marketing,they%20leave%20your%20medical%20practice. Accessed 24 Mar 2024.

[18] https://www.gmrwebteam.com/blog/1st-pillar-role-digital-marketing-improving-patient-satisfaction. Accessed 24 Mar 2024.

Conversion Rates

Conversion rates measure how effectively content prompts patients to take important actions—such as booking an appointment or signing up for a newsletter. By dividing the number of completed actions (e.g., bookings) by the number of content views or visits, this metric helps assess the real-world impact of your marketing [9].

Take for example a medical practice that that launches a chronic illness management video on their website. If the practice's landing page offers a clear call to action for scheduling a consultation, the conversion rate will reflect how effectively the content persuades viewers. A high conversion rate signals that the message resonated and motivated action.

Tracking these metrics allows healthcare marketers to test and refine content, aligning messaging more closely with what patients actually respond to.

Other Performance Metrics

Conversion isn't the only measure of success. Other key indicators, such as website traffic, social media engagement, and patient feedback, reveal what content resonates with your audience and inform strategic adjustments [10].

- **Website Traffic** shows volume and referral sources.
- **Social Media Interaction** reveals what content prompts sharing or commenting.
- **Patient Feedback** highlights patient priorities and uncovers gaps in care communication.

Together, these metrics help identify strengths and weaknesses in content strategies and guide future campaign planning.

Why It Matters

Effective content marketing informs patients, builds trust, and supports informed decision-making. By monitoring performance and responding promptly to the data, healthcare practices can continually optimize messaging and outcomes.

Close the loop by updating your plan and calendar each month based on the metrics.

Conclusion

Effective healthcare content marketing requires more than just creativity. It needs strategy, credibility, and patient-focused thinking. From infographics to podcasts, video campaigns to AI-powered content creation, each tool should have a clear goal:

to educate, engage, and empower patients while ensuring compliance and trust. When approached thoughtfully, these strategies not only boost your online presence but also strengthen your connection with the communities you serve. As digital platforms change, your messaging should adapt—being measured, responsive, and based on your patients' needs.

Appendix: Sample Content Calendar Template for Healthcare Marketing

A content calendar helps healthcare providers plan, schedule, and organize digital marketing efforts. This template ensures that campaigns stay consistent, timely, and aligned with your practice's goals.

Date	Platform	Content topic	Target audience	Call to action	Status
10 August 2025	**Facebook**	Back-to-school vaccination tips	Parents	Book now	Scheduled
12 August 2025	**Email**	Managing hypertension: What patients need to know	Adults 40+	Read more	Drafted
14 August 2025	**Blog**	Benefits of telemedicine for follow-up visits	General	Schedule appointment	Published
16 August 2025	**Instagram**	Behind-the-scenes at our clinic	All followers	Follow us	In progress

References

1. Spicer JO, Goldman CG. Creating effective infographics and visual abstracts to disseminate research and facilitate medical education on social media. Clin Infect Dis. 2022;74(Supplement_3):e414–22. https://doi.org/10.1093/cid/ciac051.
2. Balkac M, Ergun E. Role of infographics in healthcare. Chin Med J (Engl). 2018;131(20):2514–7. https://doi.org/10.4103/0366-6999.243569.
3. Balls-Berry J, Sinicrope P, Valdez Soto M, Brockman T, Bock M, Patten C. Linking podcasts with social media to promote community health and medical research: feasibility study. JMIR Form Res. 2018;2(2):e10025. https://doi.org/10.2196/10025.
4. Nuadox. DocsGPT: HIPAA-compliant AI for automating healthcare documentation. Nuadox (2023, April 5)
5. Globenewswire. https://www.globenewswire.com/fr/news-release/2025/01/20/3011911/0/en/Healthcare-Chatbots-Market-is-Set-to-Surpass-Valuation-of-US-11-8-Billion-By-2033-Astute-Analytica.html. Accessed 5 Aug 2025.
6. MOBI Health News. https://www.mobihealthnews.com/news/gyant-hauls-136m-series-ai-care-coordination-tool. Accessed 5 Aug 2025
7. Afzal M, French KE, Bilbrey LE, Faruki AA. Artificial intelligence in the clinic: creating harmony or just adding noise? Am Soc Clin Oncol Educ Book. 2025;45(3) https://doi.org/10.1200/EDBK-25-481490.

8. Imaging Technology News. https://www.itnonline.com/content/nuance-launches-dragon-ambient-experience-telehealth?utm_source=chatgpt.com. Accessed 5 Aug 2025
9. Pasaribu SB, Novitasari D, Goestjahjanti FS, Hendratono T. The impact and challenges of digital marketing in the health care industry during the digital era and the COVID-19 pandemic. Front Public Health. 2022;10:969523. https://doi.org/10.3389/fpubh.2022.969523.
10. Ventola CL. Social media and health care professionals: benefits, risks, and best practices. Pharm Ther. 2014;39(7):491.

Chapter 15
The Role of Influencer Marketing in Healthcare

A healthcare influencer is an individual who has established credibility and a substantial following within the healthcare sector, often through social media, blogs, or other digital platforms.

These influencers share their knowledge, experiences, and insights on health-related topics, ranging from medical advice and wellness tips to patient experiences and healthcare innovations. By their nature, influencers create content for audiences with shared interests, and their endorsements or discussions about healthcare products, services, or practices carry a level of credibility and transparency that can influence their followers' healthcare decisions and perspectives.[1,2,3,4]

They can be healthcare professionals, such as doctors and nurses, content creators focusing on health and wellness, or patients who share their journeys with specific conditions or treatments. Although they may be licensed professionals or patients with hard-won experience, influencer posts must not diagnose or treat individuals, provide personalized medical advice, or attempt to replace a visit to a clinician.

With around 302 million social media users in the United States, healthcare influencers can tap into a vast audience, making them highly effective in promoting healthcare [1].

With 92% of consumers trusting influencer content over what they see in paid ads, tapping an influencer to produce content is the most effective way to drive real

[1] https://creation.co/pharma-101/medical-influencers/. Accessed 3/23/24.

[2] https://www.forbes.com/sites/forbescommunicationscouncil/2024/01/03/the-intersection-of-social-influencers-and-healthcare/. Accessed 3/23/24.

[3] https://www.linkedin.com/pulse/healthcare-influencer-ecosystem-who-know-how-engage-dr-andr%C3%A9e-bates/. Accessed 3/23/24.

[4] https://www.medpro.com/rise-and-risks-of-healthcare-influencers. Accessed 3/23/24.

C. Quinn, *The Digital Practice*, https://doi.org/10.1007/978-3-032-07745-5_15

results.[5] Indeed, the role of influencer marketing is emerging as a significant strategy in healthcare digital marketing. This approach involves identifying and collaborating with influencers within the healthcare space. Influencers, who are often healthcare professionals, patients with considerable followings, or thought leaders, possess the trust and attention of your target audience. Their endorsement can bridge the gap between healthcare providers and potential patients or clients in a way traditional advertising cannot.

Imagine a healthcare professional specializing in diabetes management. This professional collaborates with a well-respected health influencer who has a significant following of individuals interested in managing diabetes through lifestyle changes. The influencer creates a series of posts and videos showcasing how they incorporate a new, innovative blood glucose monitoring device into their daily routine. They share personal stories about the ease of use, how it helps them make better dietary choices, and the positive impact it has on their health.

This authentic content, coming from a trusted source, resonates deeply with the influencer's audience. It not only educates them on the benefits of the device but also demonstrates its real-world application, making it far more compelling than traditional advertisements. The influencer's endorsement, rooted in genuine experience and trust, effectively drives interest and trust in the product among a community already engaged in diabetes management.[6,7]

Effectively executed influencer marketing in the healthcare sector can enhance the spread of health-related information, increase awareness of different health conditions and treatments, and motivate consumers to engage in preventive health behaviors.

Identifying the right influencers requires research. Look for individuals who align with your healthcare brand's values and mission. Their audience should mirror your target demographic, ensuring that their endorsement reaches those most likely to engage with your services.

Table 15.1 summarizes influencer types, strengths, risks, and ideal use cases for healthcare campaigns.

Collaboration can take many forms, from sponsored content and social media takeovers to guest blogging on your website. The key is to maintain authenticity; endorsements should feel genuine and reflect the influencer's real views on your services or products.

Engaging with influencers in the healthcare space not only amplifies your message but also leverages the trust they've built with their audience. This trust is paramount in healthcare, where decisions often have significant personal implications.

---

[5] https://www.forbes.com/sites/deborahweinswig/2016/10/05/influencers-are-the-new-brands/?sh=6e7cfcc77919. Accessed 3/23/24.

[6] https://www.dmnews.com/influencer-marketing-pays-for-itself-8-times-over/#:~:text=According%20to%20MuseFind%2C%2092%25%20of,emotionally%20close%20to%20their%20fans. Accessed 3/23/24.

[7] https://www.linkedin.com/pulse/92-consumers-trust-influencer-more-than-advertisement-capriati/. Accessed 3/23/24.

Table 15.1 Types of healthcare influencers and their strategic roles

Influencer type	Examples	Strengths	Risks/limitations	Ideal use case
Licensed healthcare professionals	Physicians, nurses, pharmacists	High trust, clinical authority	Must comply with HIPAA, FTC, and FDA rules	Promoting medical services or regulated health products
Wellness creators (licensed)	Registered dietitians, certified fitness trainers	Evidence-based content, niche expertise	Still requires compliance oversight when promoting health products	Preventive care, nutrition, fitness campaigns
Wellness creators (unlicensed)	Lifestyle influencers, fitness coaches	Broad reach, high engagement	May promote unregulated or unproven methods	General wellness branding or awareness campaigns
Patient advocates	Patients with chronic illness, condition bloggers	Emotionally relatable, experience-driven	May blur the line between story and medical advice	Promoting treatment options or support resources
Medical thought leaders	Academic MDs, researchers, keynote speakers	Policy influence, scientific credibility	Less accessible tone, limited social media engagement	Vaccine education or health policy communication
Health organizations as influencers	Clinics, hospitals, nonprofits	Institutional trust, professional branding	Slower engagement, may lack a personal touch	Service awareness, institutional reputation building

By carefully selecting and collaborating with influencers, healthcare providers can effectively communicate their value proposition to a broader audience, enhancing their digital marketing strategy.

Checklist: How to Vet a Healthcare Influencer

- **Disclosures.** Require clear labels such as #ad or "Paid partnership with…".
- **Evidence.** Ask for sources behind any health claim. Require medical review before posting.
- **Audience fit**. Confirm age, location, language, and condition match your target patients.
- **Real engagement.** Check 90-day engagement rate, comment quality, and signs of fake followers. Signs of fake followers include sudden follower spikes without a clear trigger, accounts with no photo or posts, generic emoji-only comments, and audience locations that do not match the influencer's stated market.
- **Content history**. Scan past posts for misleading claims, off-label promotion, or non-compliance.
- **Credentials.** If they present as a clinician, verify license, board status, NPI, and any sanctions on OIG or state boards.

(continued)

- **Conflicts.** Ask about paid relationships with competitors, device makers, or pharmacies.
- **Brand safety**. Review tone, language, and visuals for professionalism.
- **Regulatory literacy**. Confirm they understand HIPAA, FDA promotional rules, and FTC disclosure rules.
- **Contract readiness.** Use pre-approved talking points, set an approval workflow, define edit rights and usage rights, and ban PHI in comments and URLs.
- **Measurement setup**. Assign unique UTM links, a dedicated landing page, and a CRM field for influencer name.
- **Risk plan.** Name an escalation contact, set a takedown time frame, and define how comments will be monitored.

Ethical Considerations and Best Practices for Influencer Partnerships

In the realm of influencer partnerships, ethical considerations and best practices form the cornerstone of a successful and reputable campaign. Key among these considerations is the imperative of disclosure. Brands and influencers must clearly communicate any commercial relationship to their audience.

Callout Box: Compliance Rules for Influencer Posts
Use these rules for every paid or sponsored post to protect patients and stay compliant.

- Always disclose sponsorship in the first line using #ad or the platform's Paid partnership with [Brand] label.
- Do not publish testimonials that imply results or guarantee outcomes.
- Do not use before and after images for regulated drugs or devices unless labeling rules allow them and required risk information appears with the post.
- Require brand medical and legal review, with written approval, for all medical statements before posting.
- For UTM links and landing pages, never include protected health information in URLs, form fields, or comments.

This transparency is not only a matter of ethical responsibility but also a legal requirement in many jurisdictions. Influencers should explicitly state when content is sponsored, using clear and understandable language or hashtags such as #ad or #sponsored.[8]

Another critical aspect is the genuineness of the endorsement. Influencers should partner with brands that align with their values and interests, ensuring that any promoted product or service is one they genuinely use or believe in. This openness fosters trust with their audience and upholds the influencer's credibility.[9,10]

Finally, when operating within specialized fields such as the pharmaceutical industry, additional ethical considerations come into play. It is crucial to navigate these partnerships with a heightened sense of responsibility, ensuring that all content is accurate, compliant with regulatory standards, and clearly distinguishes between medical advice and personal opinion.[11]

Risks Associated with Influencer Partnerships

The integration of medical influencers into digital marketing campaigns presents a nuanced landscape for healthcare organizations. While the approach offers a dynamic avenue to connect with a broader audience, it also introduces a spectrum of risks that could potentially harm the organization's reputation and patient trust.

Callout Box: Red Flags to Spot Before You Post

- Unsubstantiated health claims or no sources.
- Off-label promotion of drugs or devices.
- Undisclosed payment, gifts, or free products.
- Testimonials that imply outcomes or guarantees.
- Before-and-after images where rules prohibit them or without the required risk info.
- Advice that looks like a diagnosis or individualized treatment.
- Claims that conflict with guidelines or FDA labeling.
- Sensational, clickbait headlines designed to game the algorithm.
- Conflicts of interest not disclosed.
- Signs of fake followers, sudden follower spikes, empty profiles, emoji-only comments, or audience locations that do not match the market.

---

[8] https://rjmtechnologies.com/ethical-considerations-in-influencer-marketing/. Accessed 3/23/24.

[9] https://www.linkedin.com/pulse/does-your-influencer-strategy-align-ethical-standards-apoorv-mathur-qoy7f/. Accessed 3/23/24.

[10] https://www.researchgate.net/publication/370158372_Beyond_the_Hype_Navigating_the_Ethics_of_Influencer_Marketing. Accessed 3/23/24.

[11] https://www.pharma-mkting.com/blog/influencer-marketing-in-pharma-ethical-considerations-and-best-practices/. Accessed 3/23/24.

The risk of spreading low-quality health information is significant, especially when influencers inadvertently share incorrect health advice, leading to widespread misinformation. Such situations can mislead the public and tarnish the reputation of healthcare organizations dedicated to evidence-based practices.

Research by Suarez-Lledo and Alvarez-Galvez has highlighted the extensive issue of health misinformation on social media, including false information about critical public health issues such as vaccines and diseases [2]. Their findings reveal that social media serves as a fertile ground for the swift dissemination of false health claims, described as "social contagion."

This phenomenon implies that misinformation spreads through social media interactions, such as shares, likes, and comments, much like a virus, thus broadening the scope and effect of misleading health information. The researchers stress the severe consequences of this issue on public health, which could result in dangerous behaviors and a disregard for scientifically validated health practices and treatments.[12]

Moreover, the professional image of healthcare organizations is at stake. If an influencer's content is deemed unprofessional or overly biased toward certain treatments without solid backing, it could tarnish the organization's image. Influencers with medical backgrounds may unintentionally amplify the risk if they promote treatments without scientific support, as discussed by Marocolo et al., who found that even academically credentialed influencers could spread misinformation [3].

These authors found that almost 40% of Instagram influencers were offering poor-quality information and only 2.7% cited a reference endorsing the information. Interestingly, they stated that the higher-quality information was not directly associated with a higher educational qualification of the influencers.

The encouragement of self-diagnosis and self-treatment represents another peril. Influencers, through their relatable narratives, might inadvertently lead individuals to bypass professional consultations, fostering a culture of self-diagnosis that could result in adverse health outcomes. This danger is particularly pronounced in the realm of social media, where the line between professional advice and personal opinion often blurs.

Lastly, the issue of trust and credibility cannot be overstated. Healthcare organizations must meticulously vet influencers to ensure their values align with those of the organization. Any discrepancy in this alignment could lead to a loss of credibility, not just for the influencer but for the organization as well.

While leveraging medical influencers in digital marketing campaigns can be a potent strategy for healthcare organizations, it necessitates a balanced approach that carefully considers the potential risks. Ensuring the accuracy of information, maintaining professional integrity, upholding ethical standards, and preserving trust are essential to mitigate these risks and harness the full potential of influencer marketing in healthcare.

A clear rapid response plan ensures that any harmful or noncompliant content is identified, escalated, and resolved quickly to protect patients and maintain regulatory compliance.

See Appendix 15.1 for a rapid response plan worksheet to document incidents, actions, and follow-up.

As shown in Fig. 15.1, the rapid response workflow runs from detection to documentation and follow-up.

[12] https://www.bu.edu/articles/2024/health-misinformation-rampant-on-social-media/

Sidebar: Rapid Response Plan for Risky Posts and Comments

- **Owner.** Name one person to monitor comments hourly for the first 48 h, then daily.
- **Alerts**. Turn on notifications and keyword alerts for product names, side effects, and your clinic name.
- **Triage.** Label each issue as misinformation, adverse event, privacy breach, complaint, or abuse.
- **Remove**. Hide or delete posts that violate policy, contain PHI, or promote unsafe acts. Save screenshots first.
- **Correct.** Post a short correction with a source, or link to your clinical page.
- **Escalate**. Notify compliance and legal within 1 h for privacy or safety issues.
- **Takedown**. Instruct the influencer to edit or remove risky content within a defined time window, for example, 4 h.
- **Document**. Log date, URL, screenshots, action taken, and outcome. Store in a shared folder.
- **Prepare**. Update FAQs and staff scripts after each incident. Adjust contract language if needed.

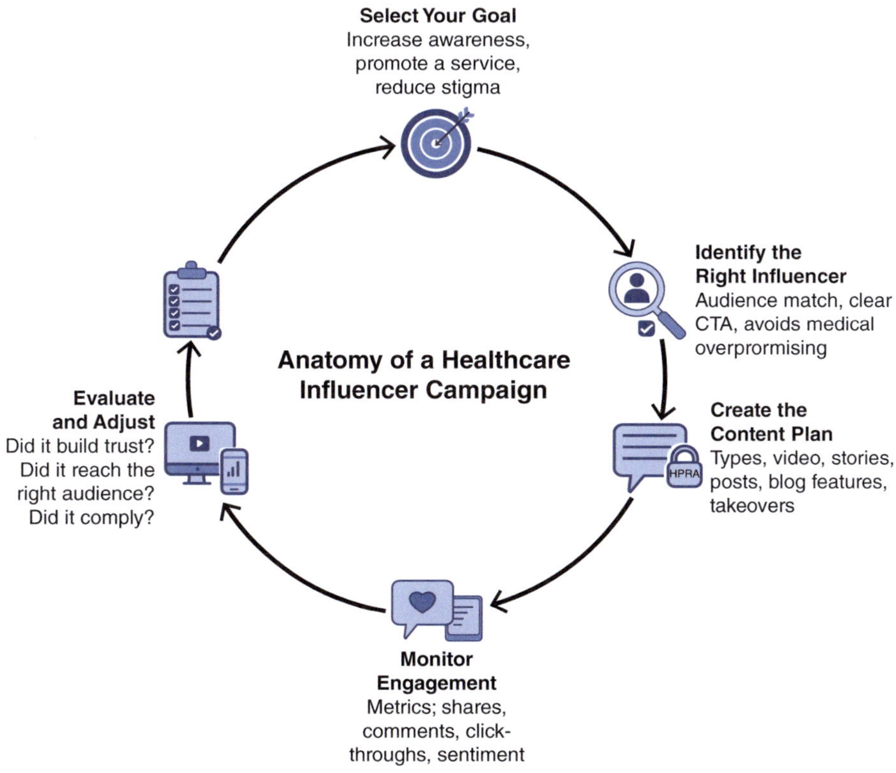

Fig. 15.1 The six essential steps of a healthcare influencer campaign, from setting goals to evaluating outcomes, with a focus on strategy, compliance, and audience engagement. (Image Quinn/ChatGPT)

Case Studies of Successful Influencer Campaigns in Healthcare

1. **Spring Health Providers Campaign**: Spring Health, a provider of mental health services, utilized healthcare influencers to raise awareness about their personalized, data-driven approach to mental health care. By partnering with influencers who shared their mental health journeys and the positive impact of Spring Health's services, the campaign effectively reached a broader audience. This approach not only humanized the brand but also helped to destigmatize mental health issues, leading to increased engagement and sign-ups.[13]
2. **Non-Opioid EXPAREL Campaign**: The marketing campaign for EXPAREL, a non-opioid pain management solution, leveraged healthcare influencers to educate the public on the benefits of non-opioid pain relief options post-surgery. Influencers, including healthcare professionals and patients who had positive experiences with EXPAREL, shared their stories across social media platforms. This strategy effectively increased awareness and trust in EXPAREL as a safe and effective pain management option, contributing to its adoption among healthcare providers and patients.[13]

Table 15.2 compares two healthcare influencer campaigns, including goals, audiences, disclosures, KPIs, results, and lessons.

Table 15.2 Influencer campaign case comparisons

Category	Spring health providers campaign	Non-opioid EXPAREL campaign
Goal	Increase awareness of personalized, data-driven mental health services and reduce stigma	Educate public and providers on benefits of non-opioid pain relief after surgery
Audience	Individuals seeking mental health support, employers, healthcare professionals	Surgical patients, healthcare providers, caregivers
Content	Influencers shared personal mental health journeys and positive experiences with spring health	Influencers (healthcare professionals and patients) shared stories and insights on EXPAREL's benefits
Disclosure method	"Paid partnership with spring health" tags and FTC-compliant disclosures	"#ad" or "sponsored by EXPAREL" tags; on-screen disclaimers in videos
KPIs	Engagement rate, number of sign-ups, referral traffic to landing page	Impressions/reach, engagement rate, provider adoption rate, patient inquiries
Results	Broader reach on social media, increased sign-ups, higher engagement	Increased awareness and adoption among providers and patients
Lesson	Authentic storytelling from influencers with lived experience builds trust and reduces stigma more effectively than purely informational content	Combining professional expertise with patient testimonials improves credibility and adoption of new medical products

[13] https://nogood.io/2023/07/05/healthcare-marketing-campaigns/. Accessed 8/2/25.

Measuring the Impact of Influencer Campaigns

You need to measure results to understand whether an influencer campaign worked. Start by identifying specific outcomes to track, then collect both numerical and qualitative feedback from each effort.

1. **Choose What to Measure**
   Match your metrics to your campaign goals. If you're trying to grow visibility, monitor reach and impressions. For engagement, focus on likes, comments, saves, and shares. To track patient interest, look at click-throughs, appointment requests, or downloads.
2. **Use Each Platform's Data**
   Ask influencers to share their content performance. Instagram shows metrics like Story taps and Saves. YouTube offers watch time and viewer retention. TikTok includes engagement breakdowns. Review the data with your team to understand which content was effective.
3. **Track Visits with Campaign Links**
   Use UTM-coded links to monitor traffic from influencer posts. Create landing pages tied to each campaign so you can track how many users visit and take action after clicking.
4. **Compare Results to Industry Norms**
   Don't evaluate performance in isolation. Engagement rates for health-related content are lower than those for lifestyle content. A 2% engagement rate may be a strong result. Use cost-per-click or cost-per-acquisition to assess value.
5. **Capture Qualitative Signals**
   Monitor comments, direct messages, and patient intake notes for signs of interest. Add influencer names to your intake forms or digital surveys. If a patient mentions an influencer, note that feedback in your customer relationship management system (CRM.)
6. **Adjust Future Campaigns**
   Use your results to improve future campaigns. If video posts attracted more clicks than images, shift your strategy. Replace underperforming formats. Focus your budget on what delivered real engagement or conversions.

Final Point

Build measurement into your campaign process from the beginning. Track both what the influencer posts and what your audience does in response. When you measure outcomes correctly, you improve both marketing performance and patient connection.

Conclusion

Influencer partnerships can help healthcare organizations reach targeted audiences, build trust, and drive engagement, but they must be approached with careful planning and compliance in mind. Success depends on selecting the right influencers, ensuring all content aligns with regulatory standards, and tracking performance against measurable goals.

Action Steps for Implementation:

- Choose the influencer type that best matches the campaign goals and audience.
- Run a full vetting checklist to verify credibility, compliance history, and audience authenticity.
- Sign a contract that includes required disclosures and content approval processes.
- Prepare a landing page that aligns with the influencer campaign and is optimized for conversions.
- Set clear KPIs to measure success.
- Schedule weekly reviews to monitor performance, address issues, and adjust tactics.

Appendix 15.1: Rapid Response Plan Worksheet

Purpose Use this worksheet to document incidents, track actions, and record follow-up measures. Each column represents a separate incident; each row captures a specific detail.

Importance and Use

This rapid response plan worksheet serves as both a compliance safeguard and a management tool. It ensures that any harmful or noncompliant content is identified, escalated, and resolved using a documented, step-by-step process. Physicians use the form to verify clinical accuracy, confirm that corrective actions meet professional and regulatory standards, and provide input on prevention strategies. Completed worksheets create a permanent record with forensic value in the event of a legal or regulatory review and offer management value by highlighting patterns, training needs, and workflow improvements.

Incident details	Incident 1—sample	Incident 2	Incident 3
Date & time incident detected	Mar 12, 2025—10:45 AM		
Incident type	Unverified medical claim in social media ad		
Detected by	Marketing coordinator during routine ad review		
Initial assessment	Ad copy stated "This procedure guarantees 100% recovery," which is unsubstantiated		
Assigned lead	Compliance Officer—Jane Smith		

(continued)

Incident details	Incident 1—sample	Incident 2	Incident 3
Actions taken	Paused campaign, notified compliance/legal, revised copy, obtained physician review		
Escalated to	Chief Compliance Officer, Medical Director		
Resolution date & time	Mar 12, 2025—2:15 PM		
Resolution summary	Ad revised and approved; relaunched with compliant language		
Follow-up actions/ prevention steps	Added compliance review step before launches; provided refresher training on prohibited claims		

References

1. Ruby D. Social media users in the world—(2023 Demographics). DemandSage. 2023, March 20. Retrieved from www.demandsage.com/social-media-users/. Accessed 23 Mar 2024.
2. Suarez-Lledo V, Alvarez-Galvez J. Prevalence of health misinformation on social media: systematic review. J Med Internet Res. 2021;23(1):e17187. https://doi.org/10.2196/17187. PMID: 33470931; PMCID: PMC7857950.
3. Marocolo M, Meireles A, de Souza HLR, Mota GR, Oranchuk DJ, Arriel RA, Leite LHR. Is social media spreading misinformation on exercise and health in Brazil? Int J Environ Res Public Health. 2021;18(22):11914. https://doi.org/10.3390/ijerph182211914. PMID: 34831671; PMCID: PMC8618405.

Chapter 16
Utilizing Patient Testimonials and Case Studies

Introduction

Patient testimonials and case studies have become essential tools in healthcare marketing. They build credibility, support informed decision-making, and reinforce a provider's reputation through real-world experiences.[1] These stories create an emotional connection that advertising alone rarely achieves.

Today, testimonials go beyond written quotes on a website. Patients now share their experiences through video, podcast interviews, and online review platforms such as Healthgrades, Zocdoc, and Google Reviews.[2] These formats serve as modern-day referrals, offering prospective patients a personal lens through which to evaluate care quality, provider competence, and overall satisfaction.

Later sections will address the legal and ethical requirements that must guide the use of patient testimonials and case studies in healthcare marketing.

Research shows that patient narratives are more likely than generic marketing content to influence decision-making because they create relatable, experience-based trust [1].[3] Case studies add another layer of persuasion by walking the reader through a clinical journey, from presenting problem to intervention to outcome.[4] When told clearly and with proper consent, these narratives can highlight a provider's problem-solving ability, compassion, and expertise.

---

[1] Market My Market. https://www.marketmymarket.com/using-patient-reviews-and-testimonials-in-digital-marketing/?utm_source=chatgpt.com. Accessed 8/5/25.

[2] Medical Economics. https://www.medicaleconomics.com/view/online-reviews-are-becoming-more-important-to-patients-in-choosing-their-care-how-to-manage-your-online-reputation-in-health-care?utm_source=chatgpt.com. Accessed 8/2/25.

[3] https://www.medfluenceadvisors.com/blog/utilize-patient-testimonials-in-digital-marketing-strategy/. Accessed 8/9/25.

[4] Social Targeter. https://www.socialtargeter.com/blogs/analyzing-the-impact-of-case-studies-in-healthcare-marketing-strategies?utm_source=chatgpt.com. Accessed 8/5/25.

C. Quinn, *The Digital Practice*, https://doi.org/10.1007/978-3-032-07745-5_16

Effective use of patient stories requires more than good storytelling. You must balance emotional impact with legal and ethical safeguards. This chapter explains how to collect, format, and share patient experiences while protecting privacy and complying with regulations.

To begin, let's examine the ethical considerations and consent requirements that must guide the use of testimonials in any healthcare setting.

Guidelines for Collecting and Using Patient Testimonials Ethically

Healthcare providers must explain how they plan to use patient testimonials. They should state where the testimonial will appear, how long it will remain public, and what format it will take. Patients need time to review this information and ask questions before agreeing to share their story [2].

Consent does not end once the patient signs a form. Providers must keep protecting patient privacy. They can remove or alter identifying details. They should also let patients read the final version before it goes public [3].[5]

These practices must comply with HIPAA privacy rules as well as the Federal Trade Commission's endorsement guidelines, which require that testimonials be truthful, not misleading, and representative of typical patient experiences.

Table 16.1 summarizes do's and don'ts for HIPAA-compliant patient testimonials.

To help distinguish between acceptable and unacceptable patient endorsements, the following examples show how a testimonial should be phrased to meet legal, ethical, and regulatory requirements, compared with a version that violates compliance standards. This contrast makes it clear how small changes in language can shift a testimonial from compliant to non-compliant.

Table 16.1 Do's and don'ts for using patient testimonials

Do's	Don'ts
Obtain written informed consent before using any patient story	Offer payment or rewards in exchange for a positive testimonial
Explain exactly how, where, and for how long the story will appear	Edit or alter statements in a way that changes the meaning or intent.
Remove or alter identifying details when necessary to protect privacy	Share stories without verifying that the signer is the patient or authorized representative
Allow the patient to review and approve the final version before publication	

[5] https://www.hhs.gov/hipaa/for-individuals/guidance-materials-for-consumers/index.html

Example: Compliant vs. Non-compliant Testimonial

- **Compliant:** "Dr. Smith explained my treatment options clearly, and my pain improved within 3 weeks. I appreciated her careful follow-up."
- **Non-Compliant:** "Dr. Smith's treatment guarantees 100% recovery from chronic back pain and works better than all other doctors." (Violates FTC truthfulness requirements and implies an unsupported guarantee.)

All testimonials must reflect honest experiences. Providers should avoid offering rewards for positive reviews.[6] Any attempt to influence or exaggerate a testimonial risks breaking trust and misleading potential patients [4].

Marketers should use testimonials to inform, not to manipulate. Each message should aim to educate patients about real care options. The focus should remain on accuracy and respect for the person behind the story.

By following these steps, providers can share patient stories in a way that supports marketing while protecting dignity and trust.

Figure 16.1 shows the step-by-step process for ethical collection, approval, and use of patient stories.

Fig. 16.1 A step-by-step checklist to guide healthcare providers in the ethical collection, approval, and use of patient stories in marketing. (Image Quinn/ChatGPT)

[6] https://meddkit.com/how-to-get-patient-reviews/. Accessed 8/6/25.

Crafting Compelling Case Studies

Case studies help potential patients understand what to expect from your care. These stories provide clear, relatable examples of how your team handles real clinical challenges. When patients see others with similar conditions succeed, it gives them confidence that they've found a provider who can help them too.

To create an effective case study, focus on building a concise, engaging narrative. Start by describing the patient's initial condition and the specific problems they faced. For example, you might introduce a patient who struggled with managing diabetes. Include details such as missed appointments, medication confusion, or complications from poor glycemic control.

Next, describe how your practice intervened. Highlight the clinical plan—whether it included education, digital tools, follow-up visits, or coordination with a care team. This shows your strategic approach, not just your clinical knowledge. Emphasize the role of provider empathy, communication, and tailored treatment.

Then, present outcomes that matter. Share quantifiable results like A1C improvement or medication adherence, along with the patient's own feedback. When possible, include a short quote. Make sure you get written consent, and avoid including identifiable health details.

Even when stories are anonymized or presented as composites, written consent is required if they are based on real, identifiable patient encounters.

Adjust the tone, level of medical detail, and language depending on the intended audience. A case study for prospective patients should emphasize clarity and reassurance, while one for professional peers can include more clinical terminology and outcome metrics.

Example

The following example is fictional and not based on any real patient. It is included to illustrate how a case study can be framed to highlight the patient's initial challenges, the interventions provided, and the resulting outcomes. By reviewing this example, readers can see how to structure their own case studies in a way that is clear, engaging, and ethically sound.

Before: Maria, a 58-year-old with type 2 diabetes, struggled with inconsistent medication use and an A1C of 9.2%. She missed several follow-up appointments and was unsure how to adjust her diet.

Intervention: The care team provided structured diabetes education, scheduled biweekly telehealth check-ins, and coordinated with a dietitian.

After: Within 4 months, Maria's A1C dropped to 7.0%, she attended all scheduled visits, and reported increased confidence in managing her condition. She described her care team as "supportive, thorough, and always available when I had questions."

Always consider ethics. Never use a patient's story without permission. If needed, change non-clinical details or use composites to protect identities.

Strong case studies reflect real care, not scripted marketing. When done right, they highlight your clinical value, build credibility, and help new patients trust your practice before they ever walk in the door.

Legal Considerations and Patient Consent

Healthcare marketing must follow HIPAA rules that protect patient privacy. Any time a provider wants to share a patient's story, they need written consent. This consent must explain what information will be shared, how it will be used, and where it will appear—for example, on the practice website, in a brochure, or on social media.

A strong consent form includes:

- A clear description of the information to be shared.
- The platforms where the story will appear.
- How long the story will be available.
- A statement that the patient can revoke consent at any time.

Before sharing a story, practices must verify that the person signing the form is the actual patient or a legally authorized representative. This step ensures the right person is making the decision.

Legal counsel should review all consent templates. Attorneys can ensure the language meets state and federal laws and can help practices avoid legal problems that might result from unclear or incomplete forms. While HIPAA establishes the federal baseline for patient privacy protections, some states have stricter laws governing the use of patient information in marketing. Providers should review applicable state regulations to ensure compliance beyond federal standards.

Consent is more than a legal requirement. It reflects a provider's respect for patient autonomy. By taking time to explain how stories will be used and giving patients control over that process, providers build trust while staying compliant.

Sidebar: FTC Guidelines for Healthcare Testimonials

The Federal Trade Commission (FTC) enforces advertising and endorsement rules that apply to healthcare marketing:

- Testimonials must be truthful, not misleading, and representative of typical patient outcomes.
- Any material connection between the provider and the patient giving the testimonial (such as compensation) must be disclosed clearly.
- Claims about outcomes must be supported by reliable evidence, not assumptions or isolated experiences.

- Disclosures must be easy to find and understand; they should not be buried in fine print or hidden in links.

Following these FTC requirements helps ensure transparency and maintains public trust in your marketing.

Patient stories offer powerful ways to build credibility, but only when shared with care. With the right legal steps in place, your team can protect privacy while still showing the real-world impact of your care.

Callout Box: Key Elements of a Patient Testimonial Consent Form
A patient consent form for marketing use should include the following elements:

- A clear explanation of the purpose of the testimonial.
- A description of the exact information to be shared (e.g., diagnosis, treatment, outcome).
- A list of media channels where the story may appear (e.g., website, social media, print ads).
- Disclosure of whether the patient's name, photo, or video will be used.
- The duration of time the testimonial will remain in use.
- A statement confirming that participation is voluntary and will not affect current or future care.
- The patient's right to withdraw consent at any time.
- Contact information for questions or revocation of consent.

By including these elements, providers ensure transparency, legal compliance, and respect for patient autonomy.

Using Patient Stories Across Digital Channels

Patient stories add emotional credibility to your brand, but their impact depends on strategic distribution. Sharing these narratives across digital platforms helps prospective patients see themselves in your care.

Website Integration
Start with your website. Create a dedicated "Patient Stories" page to house full testimonials and case studies. Add short excerpts to relevant service pages. Use internal links to guide readers to complete stories. Keep the content accessible and optimized for mobile viewing.

Social Media Adaptation
Use social media to share visually engaging versions of patient stories. Turn testimonials into graphics or short videos with captions. Share them through Instagram Stories, Facebook Reels, and LinkedIn posts. Always include a call to action, such as "Read the full story" or "Book an appointment."

Email Campaign Use
Incorporate patient testimonials into newsletters and appointment reminders. Focus on strong outcomes and relatable experiences. Link to the full story on your website. Write subject lines that emphasize patient success, not promotional language.

Video and Multimedia Deployment
When patients consent to appear on camera, produce short testimonial videos. Add them to YouTube, landing pages, and patient portals. Include captions for accessibility and upload transcripts to improve SEO.

In-Office Displays
Use digital signage to showcase patient stories in waiting rooms. Create slideshows or silent video loops that focus on results, satisfaction, and teamwork. Keep the tone informative rather than promotional.

Search and Review Profile Integration
Add story excerpts or video links to your Google Business Profile and healthcare review sites. Post updates that feature patient experiences with a link to full testimonials hosted on your website.

Key Reminders
Use consistent branding, tone, and formatting across all platforms. Make sure content stays within the scope of each patient's consent. Track engagement metrics such as clicks, shares, and form completions to evaluate effectiveness and make refinements.

Track performance using platform-specific benchmarks. For example, click-through rates on healthcare email campaigns often range from 2% to 4%, Facebook engagement rates from 0.5% to 1%, and Instagram engagement rates from 1% to 3%. Compare your results to these averages to evaluate effectiveness and identify where to refine content.

Content can be adapted for multiple channels, but the use must remain within the scope of each patient's consent. For instance, a full-length video testimonial approved for your website can be edited into a 30-s Instagram Reel only if that format was covered in the consent agreement.

If patient feedback raises concerns or a compliance risk is identified, follow the steps outlined in Appendix 15.1: Rapid Response Plan Worksheet to document the issue, take corrective action, and track resolution.

Strategically placed patient stories help potential patients feel seen, informed, and confident in your care. With the proper structure and consent, these narratives can become your most persuasive marketing assets.

Table 16.2 compares platforms for sharing patient stories, including strengths, limitations, and best uses.

Table 16.2 Platform comparison for sharing patient stories

Platform	Content type	Best use	Audience demographics	Tone and format	Key benefits	Risks and limitations
Practice website	Long-form case studies, video testimonials	Detailed storytelling and SEO	Adults 35+, professional audiences	Formal, narrative (text/video)	High control, improves search visibility, supports branding	Requires ongoing updates; limited viral potential
Facebook	Short videos, image quotes, links	General awareness and community engagement	Adults 30–65	Personal, conversational	Large user base, good for community sharing	Comment moderation required; limited reach without promotion
Instagram	Reels, image carousels, stories	Visual storytelling and trust-building	Young adults 18–45	Emotional, visual (photo/video)	High engagement; ideal for short-form visuals	Content is short-lived unless archived
YouTube	Video testimonials	Deep-dive storytelling and education	Wide range, skewing younger	Professional, polished (video)	High visibility; good for search and credibility	Requires production resources; comments need moderation
Email campaigns	Written testimonials, short videos	Direct patient outreach and follow-up	Existing patient base	Direct, personalized (text/media)	Targeted distribution, measurable click-through rates	Risk of low open rates if not segmented properly
LinkedIn	Case studies, provider spotlights	B2B branding and peer reputation	Professionals, clinicians, executives	Analytical, professional (text/video)	Builds authority with colleagues and referral sources	Limited impact for patient-facing campaigns
TikTok	Short, informal videos	Brand visibility among younger patients	Gen Z and millennials (13–35)	Casual, creative (video)	Viral potential, relatable content for digital-native users	Demands frequent updates; less control over sharing and trends

Conclusion

Patient testimonials and case studies help healthcare providers build trust and increase visibility. When handled with care, these narratives present real outcomes that help prospective patients feel more confident in their choices. To use them effectively, providers should explain the purpose clearly, protect patient privacy, and follow all legal requirements, including obtaining informed consent and verifying accuracy. These stories should demonstrate the provider's commitment to ethical care and honest communication. By treating testimonials as part of the patient experience rather than as promotional content, marketers uphold the same standards expected in clinical practice and strengthen the credibility of their brand.

Implementation Checklist for Integrating Testimonials into Marketing Campaigns

- ☐ Obtain written informed consent that specifies use, duration, and distribution channels
- ☐ Ensure accuracy by confirming facts and avoiding misleading claims
- ☐ Remove or alter identifying details to protect patient privacy
- ☐ Adapt the story appropriately for each marketing channel while remaining within consent limits
- ☐ Incorporate stories into multiple platforms such as websites, social media, and email campaigns
- ☐ Review all testimonials at least quarterly to ensure accuracy and continued relevance

References

1. Eggeling M, Bientzle M, Korger S, Kimmerle J. The impact of patient narratives on medical students' perceptions of shared decision making: a randomized controlled trial. Med Educ Online. 2021;26(1):1886642. https://doi.org/10.1080/10872981.2021.1886642. Erratum in: Med Educ Online. 2021 Dec;26(1):1893056. https://doi.org/10.1080/10872981.2021.1893056. PMID: 33588696; PMCID: PMC7894447.
2. Sil A, Das NK. Informed consent process: foundation of the researcher-participant bond. Indian J Dermatol. 2017;62(4):380–6. https://doi.org/10.4103/ijd.IJD_272_17. PMID: 28794585; PMCID: PMC5522778.
3. Mertens S. The check-up before publication: once the peer-review process has been completed, scientific texts are revised one final time by the editors to make them easier to read and to understand. Dtsch Arztebl Int. 2008;105(51–52):897–9. https://doi.org/10.3238/arztebl.2008.0897. Epub 2008 Dec 22. PMID: 19561804; PMCID: PMC2689626.
4. Bossema FG, Burger P, Bratton L, Challenger A, Adams RC, Sumner P, Schat J, Numans ME, Smeets I. Expert quotes and exaggeration in health news: a retrospective quantitative content analysis. Wellcome Open Res. 2019;4:56. https://doi.org/10.12688/wellcomeopenres.15147.2. PMID: 31346551; PMCID: PMC6619368.

Chapter 17
Navigating the Challenges of Healthcare Advertising on Social Media

Introduction

Social media now plays a key role in healthcare marketing. According to Hootsuite, 82% of U.S. users aged 12 and older use social media, highlighting its reach across broad demographics.[1] Platforms such as Facebook, Instagram, and TikTok offer direct impact on patient outreach and engagement. One dermatology clinic that embraced TikTok content and quickly gained visibility illustrates the trend of "dermfluencing," where providers educate while building patient interest—these efforts have become common enough to drive significant attention and clinic traffic.[2]

From the outset, healthcare advertisers must balance the opportunities of social media with the challenges of regulatory compliance and the need to protect patient trust. Selecting the right platforms and tailoring messages to each audience is essential for maximizing engagement and ensuring a strong return on investment.

Despite this potential, healthcare providers face a dynamic platform environment and strict regulatory requirements. This chapter provides actionable guidance to help your team choose effective social platforms, tailor content correctly, engage ethically, and measure results, while protecting patient privacy and maintaining credibility.

---

[1] Hootsuite. https://blog.hootsuite.com/social-media-demographics/?utm_source=chatgpt.com. Accessed 8/6/25.

[2] Allure. https://www.allure.com/story/dermatologists-and-social-media?utm_source=chatgpt.com. Accessed 8/6/25.

C. Quinn, *The Digital Practice*, https://doi.org/10.1007/978-3-032-07745-5_17

Strategies for Engagement

When using social media for healthcare advertising, understanding and engaging with diverse demographics across various platforms is paramount. Each social media platform hosts a unique audience with distinct preferences and behaviors. Tailoring content to meet these specific needs and preferences can dramatically enhance the effectiveness of advertising campaigns.

Table 17.1 summarizes platform audiences, preferred content, and primary use in healthcare marketing.

Tailoring Content for Platform-Specific Audiences Each social media platform has its own culture and user expectations. For instance, LinkedIn users seek professional and educational content, while Instagram users are drawn to visually appealing posts and stories. Healthcare advertisers must craft their messages to resonate with the audience on each platform. This involves not only choosing the right type of content (e.g., articles, videos, infographics) but also adapting the tone and style to fit the platform's norms.[3]

Understanding Demographics and Preferences Demographic analysis is crucial for effective engagement. Utilizing the analytics tools provided by social media platforms can offer insights into the age, location, interests, and behavior of your audience.

This data allows healthcare advertisers to tailor their content more precisely, ensuring it speaks directly to the audience's needs and preferences. For example, younger demographics might prefer engaging, short-form videos on TikTok or

Table 17.1 Social media platforms—audience, content, and use in healthcare marketing

Platform	Dominant age group	Preferred content type	Key strengths in healthcare marketing
Facebook	35–64	Articles, videos, patient stories	Broad reach, trusted by older demographics
Instagram	18–34	Images, short-form videos, reels	Visual storytelling, strong for wellness and aesthetics
TikTok	16–29	Short-form, casual video content	High engagement with Gen Z, ideal for education and myth-busting
LinkedIn	30–55 (professionals)	Articles, infographics, thought pieces	Best for B2B marketing, recruiting, and building credibility
YouTube	18–49 (broad range)	Long-form and short-form videos	High trust, effective for explainer videos and patient education
Twitter/X	25–49	Threads, real-time updates	Useful for live event coverage and professional dialogue

[3] https://sproutsocial.com/insights/audience-engagement/. Accessed 3/25/24.

Instagram, while older audiences might appreciate detailed articles or informative posts on Facebook or LinkedIn.[4]

Maintain consistent branding elements such as logo placement, color schemes, and tone of voice across all platforms to reinforce recognition and trust, even when content formats differ. Healthcare organizations typically see strong performance on Instagram, with average engagement rates of 3.7% for standard posts and 2.7% for Reels, well above general platform benchmarks. On Facebook, healthcare posts average 1.9% engagement.[5] Facebook ad click-through rates in healthcare land close to 0.83%, a helpful reference point when evaluating content effectiveness. When adapting content for multiple platforms, confirm that all formats and uses are covered under the patient's original consent agreement to prevent compliance issues.

Creating Engaging and Effective Content Engagement is key to successful social media advertising. Content that sparks conversation, encourages sharing, and invites participation can significantly increase visibility and impact.

Techniques such as asking questions, creating polls, and encouraging user-generated content can foster a sense of community and engagement. Additionally, breaking down complex healthcare information into digestible, relatable content can help demystify healthcare topics and promote informed decisions.[6] Figure 17.1 shows the best content types to post on each platform and the key metrics to track.

[4] https://www.linkedin.com/pulse/using-demographics-help-tailor-your-social-media-posts-brianna-moyer/. Accessed 8/9/25.

[5] https://blog.hootsuite.com/healthcare-social-media-benchmarks/?utm_source=chatgpt.com. Accessed 8/9/25.

[6] https://help.hootsuite.com/hc/en-us/articles/4403597090459-Create-engaging-and-effective-social-media-content. Accessed 8/9/25.

Fig. 17.1 A platform-by-platform guide for healthcare marketers, showing the best content types to post and the key performance metrics to track on Facebook, Instagram, TikTok, LinkedIn, YouTube, and Twitter/X. (Image Quinn/ChatGPT)

Social Media in Healthcare: What to Post, Where to Post It, and What to Watch

Platform	Best Content Types	Key Metrics to Track
Facebook	Patient stories, clinic updates health articles	Engagement rate (avg. 1.9%), CTR, reach
Instagram	Visual wellness tips, infographics Reels	Engagement rate (avg. 3.7%), saves, shares
TikTok	Short Q&A videos, myth-busting, tutorials	Video views, comments, engagement (1.0%)
YouTube	Professional insights, hiring posts, infographics	Click-throughs, shares, impressions
Twitter/X	Explainer videos, how-tos, patient education	Watch time, subscribers, conversion rate

Regulatory Framework for Healthcare Advertising on Social Media

Advertising healthcare services on social media requires careful attention to regulations that protect patient privacy and prevent misinformation. The Health Insurance Portability and Accountability Act (HIPAA) sets the legal foundation. It requires healthcare organizations to keep patient information confidential and secure in all marketing communications. While HIPAA provides the federal baseline for patient privacy protections, some states impose stricter rules governing the use of patient information in marketing. Healthcare organizations should review both federal and state requirements before launching campaigns.

Sidebar: FTC Endorsement Guidelines for Healthcare Advertising

- Testimonials must be truthful and accurately represent actual patient experiences.
- Any material connection between the advertiser and the endorser (such as payment, discounts, or gifts) must be disclosed clearly.
- Statements about outcomes must be supported by reliable, scientific evidence.
- If the results shared are not typical, this must be clearly stated (e.g., "Results not typical. Your experience may vary.").
- Visual content, such as before-and-after photos, must be authentic and not digitally altered in a way that misleads viewers.
- Disclosures must use plain language, be easy to read, and be placed where viewers will see them without extra clicks.
- On social media, disclosures should appear in captions or tags at the beginning of the post (e.g., #ad, "Paid Partnership with…") and remain visible for the entire viewing.

Example:

- **Compliant:** "I lost 15 pounds in 3 months using Dr. Smith's program. Results not typical. Paid partnership."
- **Non-compliant:** "Best weight loss ever!" without any disclosure of payment or clarification of typical results.

HIPAA Compliance in Social Media Advertising

Healthcare providers must obtain written authorization before using any patient information in advertising. That authorization must clearly state what data will be used, how it will be shared, and where it will appear. Providers must also apply safeguards that limit unauthorized access, monitor data handling, and ensure all content protects privacy throughout the marketing process.

Avoiding Misleading Claims

Healthcare organizations must present accurate, evidence-based information in all advertising. Credible sources and clinical data must support any claims about treatments, outcomes, or benefits. Overstating results or implying guaranteed success can mislead patients, violate advertising laws, and damage professional credibility. Misleading claims not only risk regulatory action but also erode patient trust, which is essential for sustaining long-term relationships.

For example, in 2018, the Federal Trade Commission settled with **Roca Labs**, requiring them to reimburse over $409,000 to consumers.[7] The company marketed a dietary supplement as a gastric bypass alternative, used testimonials and pricing schemes to exaggerate effectiveness, and imposed gag rules on unhappy customers. Regulators found the lab's claims lacking scientific support and deceptive, resulting in enforcement action and refunds.

Healthcare advertising on social media requires more than compliance. It calls for responsible communication that protects patients and upholds the reputation of the organization. When providers respect privacy laws and avoid exaggerated claims, they maintain public trust and ensure their marketing aligns with legal and ethical standards.

Sidebar: Common Types of Misleading Claims to Avoid in Healthcare Marketing

Type of claim	Why it's misleading
Guaranteed results	No treatment works 100% of the time. Claims like "guaranteed weight loss" violate truth-in-advertising standards
Cure language	Phrases like "cures cancer" or "eliminates diabetes" overstate medical capabilities and raise regulatory scrutiny
Before-and-after photos without disclosure	Edited or unverified images may misrepresent outcomes and mislead consumers
Unsubstantiated testimonials	Patient stories must reflect typical results. Claims like "I lost 30 lbs. in 2 weeks" require evidence
"Clinically proven" without evidence	This phrase must be backed by published, peer-reviewed studies—Ideally with FDA or peer validation
Use of medical titles in ads without oversight	Referring to someone as "Dr." in an ad implies licensed oversight. Titles must be accurate and not misleading
Comparative claims without data	Statements like "better than surgery" or "more effective than medication" require direct head-to-head data

Tip: Always review claims with legal or compliance teams before publishing. When in doubt, stick to educational language and documented outcomes

Developing Effective and Compliant Social Media Strategies

Social media provides healthcare providers with a direct channel to educate patients and build trust. To utilize this channel effectively, providers must develop strategies that comply with regulations and target the relevant audience.

[7] FTC. https://www.ftc.gov/news-events/news/press-releases/2025/07/ftc-sends-refunds-consumers-harmed-weight-loss-supplement-marketers-deceptive-claims-review?utm_source=chatgpt.com

Understand Your Audience

Start by identifying your audience's demographics, health interests, and preferred platforms. Use analytics tools like Meta Business Suite, Sprout Social, or Buffer to track what types of posts get attention. These tools can show age ranges, location clusters, post performance, and follower behavior. This information helps shape content that matches patient needs.

Create Relevant, Practical Content

Once you understand who you're speaking to, tailor content that aligns with their concerns. A pediatric clinic might share animated videos on childhood nutrition or vaccine schedules. A surgical center might post recovery timelines or short videos on preparing for outpatient procedures. Content should be concise, accurate, and valuable. Use infographics, patient stories (with consent), or Q&A posts to simplify complex topics and prompt interaction.

Protect Patient Privacy and Ensure Accuracy

Every post must comply with privacy laws, including HIPAA. Never include patient information or identifiers without written consent. Check all health claims against peer-reviewed sources or professional guidelines. Avoid promotional language that exaggerates treatment results or guarantees outcomes. To avoid liability, do not respond to online reviews with any patient information, even if the patient disclosed it.

Include a rapid response protocol for addressing negative or non-compliant feedback. This plan should designate who monitors comments, who reviews potential compliance issues, and how escalation to legal or compliance teams is handled. See Appendix 15.1: Rapid Response Plan Worksheet for a structured template.

Engage Thoughtfully

Effective strategies include more than content creation. Engagement matters. Reply to comments, host live Q&A sessions, and join trending health conversations when appropriate. Use feedback from your audience to refine future posts. If you encounter misinformation or negative feedback, respond calmly and factually. Avoid debate. When applicable, redirect users to evidence-based resources or invite them to call the office privately.

By applying these strategies, healthcare providers can meet compliance requirements while building an authentic and helpful online presence.

Table 17.2 Key performance indicators for healthcare social media campaigns

Metric	Purpose	Healthcare benchmark	Source
Engagement rate	Measures audience interaction with posts	Facebook 1.9%, Instagram 3.7%, TikTok 1.0%	Hootsuite [1]
Click-through rate	Measures how many users clicked a link or CTA	Facebook ads: ~0.9%, healthcare PPC: ~3.3%	Varos [2] Promodo [3]
Follower growth	Tracks brand visibility and audience expansion	+5–10% quarterly (goal-based)	Industry standard
Conversion rate	Measures how many users completed a desired action	Varies by campaign (appointments, form downloads)	Campaign-dependent

Measuring Success and Adjusting Strategies

To run a successful social media campaign, healthcare providers must measure what matters. Tracking the proper performance metrics allows your team to assess what's working, spot gaps, and refine strategy over time.

Tools like **Google Analytics**, **Meta Business Suite**, **Hootsuite Analytics**, and **Sprout Social** help track results in real time. These tools reveal how many users engaged with your content, clicked through to your site, or booked an appointment. For example, a campaign promoting flu shots might receive a 2.1% engagement rate, well above the healthcare average of 1.2–1.8%, indicating strong performance.

Table 17.2 lists core social KPIs, their purpose, typical healthcare benchmarks, and sources.

After reviewing performance, adjust your approach based on what the data shows. If video posts consistently outperform static images, prioritize short-form video. If engagement drops on weekends, shift your schedule to high-traffic weekdays. If your click-through rate is below 1%, revise your calls to action or targeting.

Monitoring and adapting based on data allows healthcare organizations to stay responsive, compliant, and patient-centered.

Conclusion

Navigating the complexities of healthcare advertising on social media demands a balanced approach that aligns with regulatory requirements while engaging effectively with the target audience. Through a deep understanding of the regulatory landscape, the implementation of strategic, compliant advertising practices, and the application of data-driven insights, healthcare providers can leverage the power of social media to enhance patient acquisition and deepen patient engagement. This comprehensive strategy not only ensures compliance but also capitalizes on the ability to increase patient acquisition, streamline communication, and build community trust.

Implementation Checklist for a Compliant, Effective Social Media Strategy

☐ Define your target audience and select platforms accordingly.
☐ Align all content with HIPAA, FTC, and state-level advertising requirements.
☐ Obtain and document written consent for any patient-related content.
☐ Use only evidence-based claims supported by credible sources.
☐ Monitor performance metrics regularly and adjust strategy as needed.
☐ Have a rapid response plan in place for negative or non-compliant feedback.

References

1. Hootsuite. Healthcare social media benchmarks. 2025. Retrieved from https://blog.hootsuite.com/healthcare-social-media-benchmarks/. Accessed 9 Aug 2025.
2. Varos. Facebook CTR for healthcare benchmarks. 2025. Retrieved from https://www.varos.com/benchmarks/facebook-ctr-for-healthcare. Accessed 9 Aug 2025.
3. Promodo. Healthcare digital marketing benchmarks. 2024. Retrieved from https://www.promodo.com/blog/healthcare-digital-marketing-benchmarks-2024/. Accessed 9 Aug 2025.

Chapter 18
Email Marketing Strategies for Patient Retention and Engagement

In the rapidly evolving landscape of healthcare, maintaining a strong connection with patients is more critical than ever. Email marketing can be a powerful tool in this endeavor, offering healthcare providers a direct and personalized channel to engage with their patients. It is also a very cost-efficient way to connect, delivering a 36:1 ROI, meaning for every $1 invested, practitioners recover $36 in return from their campaigns.[1]

This chapter explores the strategic use of email marketing to bolster patient retention and enhance engagement. By exploring targeted campaigns, best practices for list management, and compliance with privacy laws, we aim to equip healthcare professionals with the knowledge to effectively utilize email marketing.

From crafting personalized messages that resonate with different patient segments to leveraging analytics for continuous improvement, these strategies are designed to foster lasting relationships with patients, ensuring they feel valued and informed. Through a combination of expert insights and practical tips, this chapter will guide healthcare providers in harnessing the full potential of email marketing to achieve their patient engagement and retention goals.[2,3,4] The next sections will guide you through segment-specific subject lines, followed by strategies for personalization, maintaining list quality, optimizing timing, and using analytics to improve results.

---

[1] Litmus. https://www.litmus.com/blog/infographic-the-roi-of-email-marketing?utm_source=chatgpt.com. Accessed 8/6/25.

[2] https://www.linkedin.com/pulse/ultimate-guide-using-email-marketing-patient-retention-pandiyan-raja-zxn3c/. Assessed 3/26/24.

[3] https://medium.com/@pandiyanr05/the-ultimate-guide-to-using-email-marketing-for-patient-retention-7b87e24ca738. Assessed 3/26/24.

[4] https://www.revenuewell.com/article/dental-email-marketing-best-practices-patient-engagement-retention. Assessed 3/26/24.

C. Quinn, *The Digital Practice*, https://doi.org/10.1007/978-3-032-07745-5_18

Table 18.1 Sample subject lines and patient segments for personalized campaigns

Patient segment	Email type	Sample subject line
New patients	Welcome series	Welcome to [your clinic name]—your first visit guide
Diabetic patients	Condition management tips	3 ways to keep your blood sugar on track this week
Parents of young children	Preventive care reminders	Is it time for your child's next checkup?
Post-surgical patients	Recovery support	Your recovery timeline and tips for the week ahead
Seniors (65+)	Wellness education	Staying active after 65—4 habits that work
Patients with missed appointments	Reactivation campaign	We missed you—Ready to reschedule?
Portal-active patients	Digital engagement promotion	3 ways to use your patient portal to stay on track
Portal-inactive patients	Reactivation campaign	Your patient portal is waiting—Log in today
Chronic pain patients	Educational content	Understanding your pain triggers—a simple guide
Patients with seasonal care needs	Health maintenance	Protect your health this season—flu shots and more

Tip: Use merge tags (e.g., [First Name]) and segmentation data to dynamically customize the subject line and message for each group

To apply these strategies effectively, it's important to align your email content with specific patient groups. Table 18.1 lists example subject lines for common patient segments to guide targeted campaigns.

Once you've identified your patient segments, the next step is to personalize your messaging. Customization builds relevance and drives stronger engagement.

Crafting Personalized Email Campaigns for Different Patient Segments

In any type of marketing campaign, the personal touch is appreciated. This is especially true in healthcare, where personalizing communication can significantly enhance patient engagement and retention.[5] Always limit personalization to the minimum necessary patient data and ensure it aligns with the scope of consent provided.

Personalized email campaigns stand out as a potent tool in this context, allowing healthcare providers to communicate more effectively with different patient

[5] https://abmatic.ai/blog/personalized-marketing-for-healthcare-strategies-and-considerations. Assessed 3/26/24.

segments. This section delves into the techniques and examples of personalization that can be employed to create more impactful email campaigns.

One effective approach is to leverage patient data to customize content according to individual health interests. For example, patients with chronic conditions such as diabetes or hypertension could receive emails with tips for managing their condition, upcoming wellness events, or new research findings relevant to their health journey. This strategy not only keeps patients informed but also reinforces the provider's commitment to their well-being.

Additionally, incorporating information from previous appointments into email communications can remind patients of their healthcare journey and upcoming milestones.[6] For instance, follow-up emails after appointments could include personalized health advice, reminders for medication refills, or prompts to schedule the next visit. This approach ensures continuity of care and keeps the patient engaged with their health management plan.

Examples of effective personalization triggers include:

- **Post-visit follow-up**: Send a thank-you email with links to aftercare instructions.
- **Lab reminder**: Notify patients when a scheduled test is due or results are available in the portal.
- **Gap-in-care outreach**: Prompt patients to schedule overdue preventive screenings.

Automated triggers send messages at the right moment based on a patient's actions, status, or care needs. These event-based emails reduce manual follow-up, improve timeliness, and keep communication relevant. By responding to real-world patient milestones, they help reinforce care plans and maintain engagement between visits. Below is a micro-template for these emails.

Micro-template: Post-visit Follow-Up Email

1. **Greeting and thanks**—"Thank you for visiting [Clinic Name] today."
2. **Next steps**—"Your aftercare instructions are available in your patient portal."
3. **Call to action**—"Log in now to review your care plan and send us any questions."

Feedback from patients is another valuable source of data for personalizing email content [1]. Surveys and feedback forms can reveal patient preferences, satisfaction levels, and areas for improvement. By using this feedback to tailor email content, healthcare providers demonstrate that they value patient input and are committed to enhancing the care experience.

By implementing these personalization techniques, healthcare providers can create more engaging and relevant email campaigns, fostering a stronger relationship with patients and supporting their journey toward better health outcomes.

---

[6] https://www.benchmarkemail.com/blog/personalization/. Assessed 3/26/24.

Even the most well-written email campaigns will fail without a clean, organized, and permission-based email list. The next section explains how to build and maintain a high-performing subscriber base.

What Are the Best Practices for Email List Management?

An effective list management strategy keeps your email campaigns compliant, targeted, and deliverable. Begin by maintaining accurate subscriber records and removing outdated or invalid addresses. This sequence can feature reminder emails asking recipients to confirm their interest. If there is no response, move the contact to a suppression list to protect your sender reputation.

Avoiding Spam Folders and Improving Open Rates Internet service providers use filters to decide whether your emails land in the inbox or the spam folder. Poor list hygiene, missing authentication, or sending to unengaged recipients can lower your sender reputation and hurt deliverability. Keep your list clean, avoid spam-triggering words in subject lines, and send to engaged subscribers on a consistent schedule. Work with your IT team to set up Sender Policy Framework (SPF), DomainKeys Identified Mail (DKIM), and Domain-based Message Authentication, Reporting, and Conformance (DMARC) to confirm that your emails come from a trusted, verified source.

Data Security and Access Control Store any exported lists or spreadsheets in secure, encrypted locations. Limit access to authorized team members through role-based permissions to reduce the risk of unauthorized use or data breaches.

Consent and Unsubscribe Compliance Honor all unsubscribe requests within 10 business days, as required by law, and maintain consent logs to document when and how each subscriber opted in. These records help resolve disputes and demonstrate compliance with email regulations.

Implement Double Opt-In Subscription Processes This method confirms the subscriber's interest twice, significantly reducing spam complaints and enhancing engagement. It involves sending a confirmation email after the initial subscription request, ensuring that only interested parties receive your communications.[7]

Segment Your Email List Tailor your communications by dividing your email list into specific groups based on demographics, behavior, or preferences. This targeted approach allows for more relevant and personalized messages, increasing engagement and response rates. For instance, sending health tips to segments interested in wellness or exclusive offers to loyal customers.[8]

---

[7] https://getwpfunnels.com/email-list-management/. Assessed 3/26/24.

[8] https://www.businessnewsdaily.com/15999-best-practices-updating-email-list.html. Assessed 3/26/24.

Create Engaging and Relevant Content Your emails should offer real value to subscribers, whether it's through educational articles, exclusive deals, or updates on products/services. Content relevance is key to keeping your audience interested and engaged. For example, a healthcare provider might send health management tips to patients with chronic conditions, patient success stories, or updates about the practice [1].[9]

Figure 18.1 summarizes four list management practices: opt-in confirmation, segmentation, list cleaning, and unsubscribe access.

Implement A Welcome Sequence A series of welcome emails can effectively introduce new subscribers to your brand, establishing a positive initial interaction. This sequence could include an introduction to your services, a special offer, or a guide on what new subscribers can expect from your emails, setting a foundation for a lasting relationship.[7]

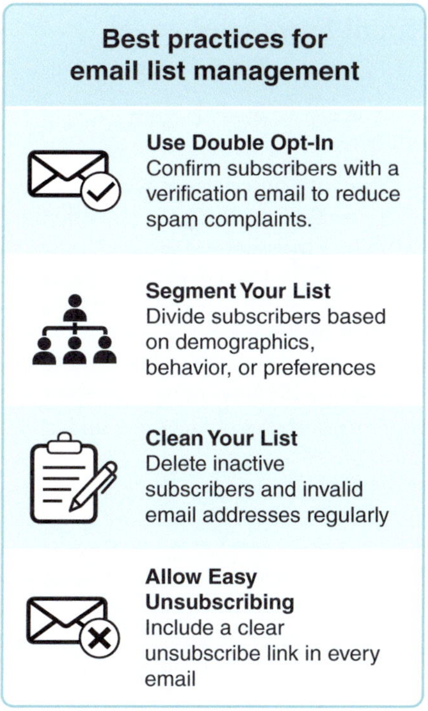

Fig. 18.1 The four essential best practices for managing email marketing lists effectively include opt-in confirmation, segmentation, list cleaning, and unsubscribe accessibility. (Image Quinn/ChatGPT)

[9] https://www.sequencehealth.com/blog/how-to-improve-your-healthcare-practices-email-marketing-strategy. Assessed 3/26/24.

Regularly Clean Your Email List To maintain high deliverability and engagement rates, it's essential to remove subscribers who no longer interact with your emails periodically. This practice keeps your list fresh and ensures your messages reach those genuinely interested.[10] Include a quarterly re-engagement sequence before removing inactive addresses. This can feature a series of reminder emails asking recipients to confirm their interest. If there is no response, move the contact to a suppression list to maintain list quality and protect sender reputation.

Once the list is built and maintained, the cadence and timing of email delivery become key factors in maximizing engagement.

Address Frequency and Timing To maintain engagement without overwhelming patients, healthcare providers should carefully manage the frequency and timing of their email communications. Both timing and frequency directly influence key performance metrics, such as open rates and click-through rates.

Email Frequency

Your email frequency should match the purpose and expectations of each patient segment:

- **Monthly newsletters** work well for general wellness updates.
- **Bi-weekly updates** are suitable for patients managing chronic conditions or following treatment plans.
- **Seasonal campaigns** (e.g., flu shot reminders) can be sent as needed.

Most healthcare audiences respond best to 1 to 2 emails per month, depending on content relevance and consent preferences.[11,12]

Along with how often you send emails, when you send them also plays a critical role in whether patients open and engage with your messages.

[10] https://www.tye.io/en/blog/email-list-management-best-practices/. Assessed 8/6/25.

[11] https://www.pdqcom.com/blog/how-often-should-i-send-hcp-marketing-communications. Accessed 8/6/25.

[12] https://www.zoho.com/campaigns/marketingmatchbox/ideal-email-frequency.html. Accessed 8/6/25.

Email Timing

Timing also plays a critical role in visibility. Industry research shows that emails sent on Tuesdays through Thursdays between 9 a.m. and 11 a.m. tend to perform best in the healthcare sector.[13] Use A/B split testing over four different send windows to identify when your audience is most responsive. Lock in the best-performing time for at least one quarter before retesting. Local time zones and the nature of each clinic's service lines can significantly affect results, so adjust benchmarks to match your audience.

Avoid sending marketing or non-urgent emails during late-night hours to prevent annoyance and reduce unsubscribe rates. For urgent communications, such as product recalls or outbreak alerts, bypass normal scheduling and send immediately to all affected recipients, using subject lines that clearly convey urgency.

To evaluate your timing and frequency choices, it helps to benchmark your email performance against current industry standards.

Healthcare Email Performance

Industry data provides helpful context for evaluating your email campaigns.[13,14,15] Figure 18.2 compares benchmark open and click-through rates from Campaign Monitor, WebFX, and 9Clouds.

If your results fall below the benchmark range, review your subject lines, audience segmentation, and send timing, then run targeted A/B tests to identify and correct weak points.

These benchmarks provide context, but seeing them applied in a real campaign illustrates how they can inform strategy.

[13] https://www.campaignmonitor.com/resources/guides/email-marketing-benchmarks/?utm_source=chatgpt.com. Accessed 8/6/25.

[14] https://www.webfx.com/blog/marketing/email-marketing-benchmarks/. Accessed 8/6/25.

[15] https://9clouds.com/blog/performance-benchmarks-for-healthcare-marketing-emails/. Accessed 8/6/25.

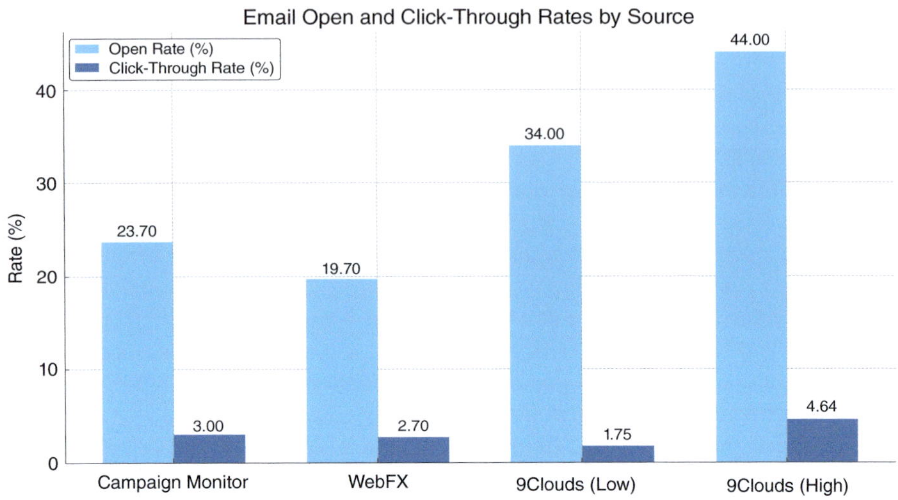

Fig. 18.2 A bar graph comparing open and click-through rates from Campaign Monitor, WebFX, and 9Clouds. It visually illustrates the performance range across healthcare email campaigns, based on credible industry benchmarks. (Image Quinn/ChatGPT)

Practical Example

A provider might send a bi-weekly email on Tuesday mornings to patients with diabetes. Each message could include seasonal nutrition tips, lab result reminders, or prompts to book follow-ups. Ensure that no lab values or other protected health information appear in the subject line or preview text. The timing and targeting help increase visibility and relevance. In addition to refining your message, test variations of the call to action and the sender name to determine which combination delivers the best response.

Pro Tip
Use A/B testing and platform analytics (e.g., Mailchimp, Constant Contact, or Klaviyo) to identify your ideal send time based on patient engagement patterns.

Incorporate Engagement Metrics Monitoring and analyzing engagement metrics are crucial for healthcare providers aiming to enhance patient engagement and retention through email marketing. Engagement metrics, such as open rates, click-through rates, and conversion rates, offer valuable insights into how patients interact with email communications. These metrics help identify what content resonates most with the target audience, allowing healthcare providers to tailor their email strategies for maximum effectiveness.[16] By incorporating analytics into patient engagement strategies, medical service providers can harness the full potential of

[16] https://fastercapital.com/topics/measuring-and-analyzing-patient-engagement-metrics.html. Assessed 3/26/24.

data-driven decisions. Analytics enable the identification of successful elements within an email campaign and areas that require improvement. This approach ensures that email content is both relevant and engaging to the patient, fostering a stronger provider–patient relationship.[17,18]

In addition to traditional metrics like open rate and click-through rate, tracking click-to-open rate (CTOR) and spam complaint rate provides deeper insight into campaign health. CTOR measures how effectively the email content motivates action among those who opened it, which helps confirm whether subject lines and body content align. A low CTOR often means the email's promise in the subject line was not delivered in the content. Spam complaint rate monitors how often recipients mark your messages as unwanted, which can damage sender reputation and lead to deliverability issues. Keeping spam complaints under 0.1% is critical in healthcare, where patient trust and compliance with privacy regulations directly affect engagement. Monitoring all metrics together allows you to diagnose performance issues and make targeted adjustments.

Table 18.2 defines key email metrics for healthcare, lists typical benchmarks, and shows actions when results fall short.

High-performing campaigns must also respect legal boundaries. Compliance builds patient trust and shields your practice from liability.

Alongside performance optimization, every campaign must meet legal and ethical standards to protect patient information and maintain trust.

Ensure Compliance with Privacy Laws Adhering to email marketing regulations, such as GDPR and CAN-SPAM, is crucial. These laws require obtaining explicit consent to send emails, providing a straightforward way to unsubscribe, and safeguarding subscriber privacy, thereby fostering trust and transparency.[19]

This journey through the intricacies of email marketing strategies for patient retention and engagement has illuminated the pivotal role of personalization, list management, and the careful calibration of frequency and timing. By leveraging patient data, healthcare providers can craft emails that resonate on a personal level, fostering a deeper connection and encouraging ongoing engagement.

Under HIPAA, marketing refers to any communication that encourages the purchase or use of a product or service not directly related to a patient's treatment. Always secure Business Associate Agreements (BAAs) with your email service provider (ESP), customer relationship management (CRM) platform, and customer data platform (CDP) before sending campaigns that involve protected health information.

---

[17] https://www.linkedin.com/pulse/why-you-should-use-analytics-improve-patient-engagement/?trk=public_post. Assessed 3/26/24.

[18] https://elitenp.com/marketing-tip-monday-77-create-effective-email-marketing-campaigns-for-patient-retention/. Assessed 3/26/24.

[19] https://flodesk.com/blog/email-list-engagement-best-practices/. Assessed 3/26/24.

Table 18.2 Key email engagement metrics and benchmarks (healthcare)

Metric	What it measures	Typical benchmark	What to watch for	Action if below benchmark
Open rate	Percentage of recipients who open the email	21.7% (Mailchimp 2023)	<15% may signal weak subject lines or timing	Test subject lines and adjust send times
Click-through rate (CTR)	Percentage of recipients who click a link	2.5–3.0%	<1% suggests content is not engaging or relevant	Improve content relevance and clarity of links
Click-to-open rate (CTOR)	Percentage of openers who click a link	10–15%	<8% may indicate that the content does not match the subject line	Align subject line and body content, improve calls to action
Unsubscribe rate	Percentage who opt out after receiving an email	<0.5%	>0.5% may indicate too frequent or irrelevant emails	Reduce send frequency or improve audience targeting
Bounce rate	Percentage of emails that fail to deliver	<2% total (soft + hard)	>2% could indicate list quality issues	Clean lists regularly and remove invalid addresses
Conversion rate	Percentage who complete a goal (e.g., appointment booked, refill completed, portal signup)	Varies (1–5% common)	Depends on the action requested	Simplify the path to completion and make CTAs prominent
Spam complaint rate	Percentage marking email as spam	<0.1%	>0.1% indicates content or targeting problems	Review audience targeting, adjust content tone, and confirm opt-in

Note: Most platforms like Mailchimp, Constant Contact, and Campaign Monitor automatically track these metrics and offer benchmarks by industry

In addition to HIPAA, state privacy laws such as the California Consumer Privacy Act (CCPA) and similar state-level regulations grant patients specific rights, including the ability to access, correct, and request deletion of their personal data. Ensure your email processes can respond to these requests promptly.

Always obtain documented consent before sending any marketing content to patients. Avoid combining treatment-related communications with marketing messages in the same email, as this can blur legal boundaries and erode trust. Maintain detailed logs showing the lawful basis for including each contact on a given marketing list.

Collect only the information required for your campaign. Use it only for the purpose stated when it was collected. Set and follow a clear retention period, deleting or anonymizing the data when it is no longer needed.

Adhering to best practices in list management and privacy laws not only ensures compliance but also builds trust with patients. Moreover, understanding the optimal frequency and timing of emails can significantly enhance the effectiveness of communication efforts without overwhelming recipients. As we've explored, these strategies are not just about maintaining a list or sending out information; they're about nurturing a relationship with patients that supports their journey toward better health outcomes.

By implementing these best practices, healthcare providers can create a more engaged patient base, ultimately contributing to a healthier community.

Quick Compliance Checklist

- ☐ Physical address in the footer
- ☐ One-click unsubscribe
- ☐ Honor opt-outs promptly
- ☐ No PHI in subject lines
- ☐ Maintain an audit trail for consent.

As data privacy rules tighten and tracking tools evolve, physicians must adapt their outreach strategies to rely on information patients share directly.

Preparing for a Cookie-Less Future: First-Party Data Strategy

Google Chrome will officially phase out third-party cookies by the end of 2025, following similar moves by Safari and Firefox. These cookies track users across websites, enabling advertisers to build detailed behavioral profiles. Due to growing concerns about privacy and user consent, regulators and browser developers are pushing toward more privacy-respecting alternatives [2].

While artificial intelligence has not caused this shift, AI-powered personalization and advertising tools have increased the demand for clean, consent-based data. For healthcare marketers, this means a fundamental change: success will depend on first-party data, that is, information that patients provide directly to your practice.

First-party data includes:

- Contact information (email, phone number)
- Appointment history and portal activity
- Website behavior (e.g., downloads or page visits)
- Responses to online forms or surveys

First-party data is obtained directly from patients via trusted channels (e.g., patient portals, direct surveys, clinical systems), giving organizations precise insight into actual patient behavior, preferences, and needs. Unlike third-party data, which is often aggregated from external, unverified sources and can be inconsistent, first-party data is typically more reliable, traceable, and current. Patient-generated health data (PGHD) collected first-hand reflects the patient's lived experience and can improve care accuracy when robust safeguards exist.[20]

To support these efforts at scale, many healthcare organizations are turning to Customer Data Platforms (CDPs).

Connecting First-Party Data to Email Strategy

First-party data powers more relevant, segmented email campaigns. For example, if a patient downloads a guide on back pain, you can trigger an automated follow-up email offering information about physical therapy options or new treatments.

To make this possible, many healthcare organizations utilize Customer Data Platforms (CDPs). CDPs aggregate data from multiple sources, including websites, intake forms, patient portals, and email platforms, and then use this data to create individual patient profiles. This enables automatic email triggers, behavior-based segmentation, and condition-specific education campaigns.[21] Most CDPs are cloud-based and hosted by third-party companies [3].

Popular CDPs used in healthcare include:

- **Salesforce Health Cloud**
- **Adobe Real-Time CDP**
- **Klaviyo** (common in multi-site practices)

These platforms store data securely in the cloud, but practices must ensure compliance with HIPAA and establish Business Associate Agreements (BAAs) with vendors.

Options for Smaller Practices

Smaller clinics that cannot invest in a full-scale CDP can still build effective first-party data strategies using:

- Secure spreadsheets for segmenting by visit type or condition.
- Email tools like Mailchimp or Constant Contact to tag and organize patients.
- Simple intake forms to gather consent, preferences, and communication goals.

[20] https://www.mmm-online.com/partnercontent/ehr-messaging-look-for-first-party-partner/. Accessed 8/6/25.

[21] https://business.adobe.com/products/real-time-customer-data-platform/rtcdp.html. Accessed 8/6/25.

Even without enterprise-level tools, smaller practices can build personalized campaigns using simpler systems.

Regardless of size, your email platform should allow you to:

- Group patients by interest or medical condition
- Personalize subject lines and message content
- Track open and click rates for engagement

Best Practices for Physicians

To build a first-party email strategy:

- Request email and communication preferences during intake and online scheduling.
- Use short forms on your website to ask about health interests or concerns.
- Select tools that allow automated, HIPAA-compliant segmentation and outreach.

As third-party tracking disappears, physicians can't rely on passive digital profiling. Instead, they must engage patients more transparently and intentionally. Using first-party data to send timely, relevant, and permission-based emails will not only improve campaign performance but also strengthen trust and compliance.

Taken together, these strategies create a robust foundation for ethical, data-driven patient outreach that improves outcomes and deepens relationships.

Conclusion

Email remains one of the most cost-effective and scalable tools for healthcare engagement, but only when used with precision and care. By segmenting patient audiences, tailoring content based on real needs, and monitoring response metrics, healthcare providers can create campaigns that educate, retain, and support patients between visits. As data privacy expectations increase and third-party tracking tools fade, success will depend on first-party data strategies that prioritize transparency and value. Whether you're running a solo practice or managing a multi-location group, effective email marketing is no longer optional; it's a clinical and operational necessity for maintaining long-term patient relationships in a digital-first healthcare environment.

References

1. Gormley J, Fager SK. Personalization of patient-provider communication across the lifespan. Top Lang Disord. 2021;41(3):249–68. PMID: 34421170; PMCID: PMC8375497.
2. Google. Privacy Sandbox: preparing for the phase-out of third-party cookies. 2024, April 23. https://privacysandbox.com/blog/phase-out-third-party-cookies-2024/
3. Gartner. Market guide for customer data platforms. 2023. https://www.gartner.com/en/documents/4012376

Chapter 19
Developing a Mobile App for Your Practice

What Is a Mobile App, and Why Do I Need One?

A mobile app is a software application designed for use on smartphones and tablets. In healthcare, mobile apps are no longer optional. They have become essential tools for enhancing patient experience, expanding access, and strengthening digital marketing strategies. With over half of U.S. patients now using mobile health apps to manage their health, these platforms have changed how people engage with medical providers [1].

Apps allow patients to schedule appointments, view medical records, set medication reminders, and send secure messages. These features promote direct communication between the patient and the practice, reduce friction in care access, and create more touchpoints for engagement. Patients increasingly expect these capabilities. For example, a recent study found that over 80% of users with a health app accessed it at least once in the past year, often for managing chronic conditions or communicating with providers [2]. Figure 19.1 shows the top features patients expect in a healthcare app and how those features support access, communication, and usability.

Mobile apps also support broader access to care. For patients in rural or underserved communities, telehealth features embedded in apps make follow-up visits and consultations more accessible. This improves continuity of care without requiring additional infrastructure. Apps can also use smartphone features such as the camera or GPS. For instance, a dermatology app might allow patients to submit photos of skin conditions for asynchronous review, or a practice app might use location services to direct users to the nearest urgent care center.

The impact on patient satisfaction and adherence is measurable. One clinic that implemented in-app reminders saw a 38% reduction in missed appointments, driven by push notifications that alerted patients to upcoming visits and encouraged rescheduling if needed [3]. This demonstrates that apps not only improve convenience but also influence patient behavior.

C. Quinn, *The Digital Practice*, https://doi.org/10.1007/978-3-032-07745-5_19

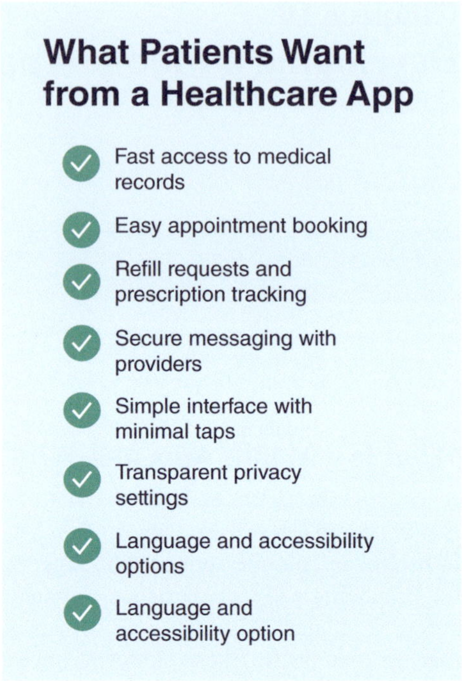

Fig. 19.1 The top features patients expect in a healthcare app to improve access, communication, and usability. (Image Quinn/ChatGPT)

Although some practices rely solely on mobile-friendly websites, apps offer advantages that websites do not. Apps provide immediate access through a single tap, unlike websites that require navigating through a browser. They also allow more personalized interactions. App-based analytics help tailor messages and recommendations to patient behavior, deepening engagement. For example, if a user downloads a physical therapy guide, the app can automatically suggest related services or prompt a follow-up appointment.

Mobile apps are also an asset from a branding perspective. They communicate a commitment to digital innovation and patient convenience. In competitive markets, offering an app can differentiate a practice and appeal to patients who prioritize accessibility and technology.

In today's mobile-first landscape, a well-designed healthcare app is more than a convenience; it is a strategic asset. It helps build stronger patient relationships, reduces administrative burden, and enhances clinical follow-up. For practices looking to meet modern expectations and improve outcomes, investing in a mobile app should be a top priority.

Table 19.1 lists core app features with plain-language descriptions for planning and building.

Incorporating these features into a healthcare professional's mobile app can significantly enhance patient care, streamline practice operations, and improve patient engagement and satisfaction. Beyond these features, mobile apps provide measurable value in clinical operations and patient behavior. The following section explores how these tools improve engagement and outcomes across a variety of care settings.

Table 19.1 Key features for a healthcare app

Feature	Description
Easy registration and login	Allows users to create accounts and log in easily, improving accessibility and user retention
Multi-language support	Users can use the language they are most comfortable communicating in
Profile access and modification	Enables users to update personal, medical, and contact information to keep records current
Appointment scheduling and management	Facilitates self-service scheduling, rescheduling, or canceling of appointments, reducing admin workload
Electronic prescribing	Supports digital prescribing directly from the app, enhancing medication safety and workflow
Secure messaging	Provides a private, secure chat feature for communication between patients and providers
Access to medical records	Grants patients access to their records, including lab results, visit summaries, and health history
Privacy, security, and regulatory compliance	Ensures compliance with HIPAA and other laws, protecting patient data and maintaining trust
Push notifications	Delivers timely reminders and health updates, encouraging adherence and proactive care

The benefits of having a dedicated mobile app for patient engagement.

The integration of mobile apps into medical practices has emerged as a transformative strategy for enhancing patient engagement. These digital tools offer a plethora of benefits that extend beyond mere convenience, significantly impacting both patient outcomes and the operational efficiency of healthcare providers.

Mobile apps have been instrumental in improving medication adherence [4]. In hypertension care, smartphone app interventions for patients with hypertension led to a significant improvement in medication adherence, with a standardized mean difference of 0.38 (95% CI: 0.26 to 0.5) [5].

By providing patients with timely reminders, educational content, and interactive features that allow for tracking their medication schedules, these apps address one of the most common challenges in patient care. The direct result is a decrease in the rates of medication non-adherence, which is a critical factor in the management of chronic conditions and overall patient health [2].

Moreover, the issue of missed appointments, which can disrupt care continuity and increase healthcare costs, is significantly mitigated by using mobile apps.[1] Appointment reminders, easy rescheduling options, and the ability to book appointments directly through the app enhance patient convenience and accountability, leading to better attendance rates and optimized scheduling for healthcare providers. A 2022 systematic review found that predictive model–based text message reminders significantly reduced outpatient no-show rates. In one randomized controlled trial, text messaging decreased no-shows by 25%, indicating strong potential for app-based reminders [6].

[1] https://www.osplabs.com/insights/how-hospital-mobile-apps-improve-patient-engagement/. Accessed 8/7/25.

Furthermore, mobile apps play a pivotal role in chronic disease management by offering personalized tools and resources that empower patients to take an active role in their health.[2] Chronic disease management refers to the ongoing care and support required to help patients manage long-term conditions like heart disease, diabetes, or asthma, often involving medication tracking, appointment adherence, and lifestyle modification.

Features such as symptom trackers, educational materials, and direct communication channels with healthcare providers enable patients to manage their conditions more effectively. This not only improves patient outcomes but also reduces the burden on healthcare systems by preventing complications and minimizing the need for emergency care.

Design alone determines whether patients will adopt and continue using an app. To turn good features into usable tools, healthcare organizations must focus on thoughtful design and user experience.

User Experience (UX) Design Principles for Healthcare Apps

Creating a healthcare app goes beyond packing in features. Successful UX design ensures the app is easy to use, accessible to all patients, and tailored to individual needs. This section outlines key UX strategies that support better patient engagement, satisfaction, and outcomes.

1. *Prioritize Accessibility*

   Healthcare apps must work for users with a wide range of physical, sensory, and cognitive abilities.[3] Designing with accessibility in mind is both a legal obligation and a commitment to patient-centered care. Standards such as WCAG 2.1 and Section 508 offer practical guidance to ensure inclusive design. *WCAG (Web Content Accessibility Guidelines) is an internationally recognized standard that outlines how to make digital content more accessible to people with disabilities* [7].

   Key features that support accessibility include:

- Adjustable font sizes and high-contrast settings for users with low vision
- Screen reader compatibility, voice controls, and keyboard-only navigation for users with mobility or visual impairments
- Offers text alternatives to images
- Simple layouts and plain language for users with cognitive or learning challenges

[2] https://www.updox.com/blog/the-role-of-mobile-apps-in-healthcare-patient-engagement/. Accessed 8/6/25.

[3] https://www.toptal.com/designers/healthcare/healthcare-app-accessibility. Accessed 8/7/25.

Incorporating these elements ensures that all patients can access, understand, and act on the information presented. Accessibility isn't an add-on. It's a core part of delivering respectful, equitable digital care.

2. *Simplify the Interface*

   Patients should be able to schedule appointments, access medical records, or refill prescriptions with minimal effort. Clean layouts, intuitive icons, and limited menu layers help users find what they need quickly. Simplicity reduces frustration and supports higher adoption across all age groups.

3. *Personalize Engagement.*

   Personalized features, like appointment reminders, daily medication alerts, or tailored health content, can increase patient involvement and satisfaction. Apps that learn user behavior and preferences are more likely to encourage long-term engagement and adherence to care plans.

4. *Include Multi-Language Support.*

   Language options are essential in diverse healthcare markets. In the United States, for example, Spanish is the primary language for over 13% of the population [8]. Offering healthcare apps in multiple languages improves comprehension, builds trust, and reduces medical misunderstandings.

Thoughtful UX design makes healthcare apps not only functional but empowering. Developers who prioritize accessibility, simplicity, personalization, and inclusive language design will create tools that improve patient experience and help providers deliver better care. In the following sections, we will highlight best practices and examples to guide development.

Security is not the only technical challenge. To be effective, apps must also connect seamlessly with other clinical tools. Integration with EHRs and practice management systems ensures the app supports, not fragments, care delivery.

Data Security and Privacy

Protecting patient information is central to trust and legal compliance in healthcare.[4] The following practices form the foundation of effective data security:

Data Encryption

Encryption safeguards sensitive patient data both at rest and in transit. It blocks unauthorized access and ensures confidentiality. This is not only a technical safeguard, but it is essential to maintaining patient safety and confidence in digital healthcare platforms. Both HIPAA in the United States and the General Data Protection Regulation (GDPR) in the European Union recommend encryption as a reasonable safeguard for protecting health information [9, 10].

[4] https://www.kiteworks.com/hipaa-compliance/guide-to-enhanced-data-security/. Accessed 8/7/25.

User Consent
Healthcare providers must obtain clear, explicit consent before collecting, using, or sharing patient information. This is a core requirement under both HIPAA and GDPR. Consent demonstrates respect for patient autonomy and reinforces trust in digital health platforms.

Anonymization of Patient Data
De-identifying or masking personally identifiable information allows providers to use health data for research and analytics without compromising privacy. GDPR specifically requires anonymization or pseudonymization when personal data is used for secondary purposes, and HIPAA outlines standards for de-identifying protected health information in accordance with privacy rules.

Even the best app will fail if patients don't know about it or can't use it easily. The next section offers practical strategies for promoting app adoption and guiding patients through onboarding.

Integration with Existing Healthcare Systems

Connecting healthcare apps with Electronic Health Records (EHRs) and practice management systems is essential for improving care delivery. Integration ensures that patient data is stored, managed, and transferred efficiently across clinical systems, reducing administrative burden and improving coordination.[5]

EHR integration enables real-time access to patient records, helping clinicians make more informed decisions. By linking EHRs with telehealth and scheduling platforms, providers can ensure data remains current, complete, and securely shared between systems.[6] This interoperability improves both care quality and workflow efficiency [11].

Integrated systems also support better billing practices and stronger care coordination. When EHRs and practice management tools operate together, staff can manage scheduling, insurance, documentation, and clinical communication from a unified interface. This approach promotes a streamlined patient journey and improves overall satisfaction [12]. [7]

Beyond convenience, integration advances data accuracy and continuity of care. It reduces duplication, delays, and errors while enabling consistent access to records across settings. These benefits are particularly valuable in telehealth models, where interoperability directly affects quality, safety, and service outcomes.

[5] https://www.nexhealth.com/resources/ehr-integration?topics. Accessed 8/7/25.

[6] https://streamlinehealth.net/emr-integration/. Accessed 8/7/25.

[7] https://www.osplabs.com/insights/how-to-integrate-ehr-in-practice-management-system/. Accessed 8/7/25.

Ultimately, the goal is to unify fragmented systems into a cohesive digital environment that supports both patients and providers. EHR integration fosters better outcomes, greater efficiency, and a more reliable care experience [12].[8]

App Promotion and Patient Onboarding

Effective promotion and onboarding are essential for encouraging patients to adopt and regularly use healthcare apps. A combination of outreach methods can improve visibility and usability:

1. **Leverage Social Media**
   Engage patients on platforms they already use.[9] Post short videos, app highlights, and patient testimonials that explain how the app improves convenience and access. Replying to comments or direct messages builds trust and increases reach.
2. **Use In-Office Promotions**
   Display posters, flyers, and QR codes in waiting areas and exam rooms. During visits, staff and clinicians should introduce the app, explain its purpose, and demonstrate how it can simplify scheduling or communication. Personal recommendations during clinical encounters have been shown to increase adoption rates [13].
3. **Host Patient Training Sessions**
   Offer brief, guided sessions, either in person or online, to walk users through the app's features. These sessions can be promoted through social media, email newsletters, and signage in your practice. Support is especially important for older patients or those less familiar with digital tools.[23]

By combining digital outreach with personal engagement, providers can increase app downloads and help patients use digital tools confidently. This improves long-term adoption and supports better care continuity.

To increase adoption of your healthcare app, it's important to choose the right mix of promotion channels based on your patient population. Table 19.2 lists top promotion channels, ideal audiences, and tips for effective onboarding.

[8] https://www.uptech.team/blog/ehr-integration. Accessed 8/7/25.

[9] https://www.appdemovideos.com/effective-marketing-strategies-for-your-healthcare-app-a-comprehensive-guide/. Accessed 8/7/25.

Table 19.2 Top app promotion channels and when to use them

Promotion channel	Best for	Tips for success
Social media	Younger or tech-savvy patients	Use short videos, testimonials, and regular posts
In-office posters & QR codes	All patients, especially walk-ins	Place near check-in, waiting areas, and exam rooms
Email campaigns	Existing patient base	Include direct download links and brief tutorials
SMS/text alerts	Patients with upcoming visits	Use opt-in messages with clear calls to action
Patient portals	Portal users already using digital tools	Add banners or pop-ups during login or check-in
Waiting room video loops	All patients, especially older adults	Highlight app benefits with captions and narration

Conclusion

Mobile apps have become a vital tool in digital healthcare marketing, enabling providers to communicate with patients, manage workflows, and support care delivery more efficiently. When well-designed, these apps improve patient engagement and protect sensitive data. Integration with existing health systems further enhances their value, reducing administrative friction and facilitating coordination across teams.

To ensure successful adoption, healthcare organizations must focus on usability, promotion, and straightforward patient onboarding. Strong user experience and outreach efforts increase both downloads and long-term use.

As digital tools evolve, healthcare apps will continue to play a central role in improving access, supporting care, and strengthening relationships between providers and patients.

References

1. Office of the National Coordinator for Health Information Technology. Individuals' access and use of patient portals and smartphone health apps: 2020. U.S. Department of Health and Human Services; 2021. https://www.healthit.gov/data/data-briefs/individuals-access-and-use-patient-portals-and-smartphone-health-apps-2020. Accessed 6 Aug 2025.
2. Krebs P, Duncan DT. Health app use among us mobile phone owners: a national survey. JMIR Mhealth Uhealth. 2015;3(4):e101. https://doi.org/10.2196/mhealth.4924. PMID: 26537656; PMCID: PMC4704953.
3. Klara. Text message appointment reminders reduce no-shows by 38%, study finds. 2023. https://www.klara.com/blog/text-message-appointment-reminders-reduce-no-shows-by-38-study-finds. Accessed 6 Aug 2025.
4. Dayer L, Heldenbrand S, Anderson P, Gubbins PO, Martin BC. Smartphone medication adherence apps: potential benefits to patients and providers. J Am Pharm Assoc. 2013;53(2):172–81. https://doi.org/10.1331/JAPhA.2013.12202. PMID: 23571625; PMCID: PMC3919626.

5. Xu H, Long H. The effect of smartphone app–based interventions for patients with hypertension: systematic review and meta-analysis. JMIR Mhealth Uhealth. 2020;8(10):e21759. https://doi.org/10.2196/21759. URL: https://mhealth.jmir.org/2020/10/e21759
6. Oikonomidi T, Norman G, McGarrigle L, Stokes J, van der Veer SN, Dowding D. Predictive model-based interventions to reduce outpatient no-shows: a rapid systematic review. J Am Med Inform Assoc. 2023;30(3):559–69. https://doi.org/10.1093/jamia/ocac242. PMID: 36508503; PMCID: PMC9933067.
7. World Wide Web Consortium (W3C). Web content accessibility guidelines (WCAG) 2.1. 2018. https://www.w3.org/TR/WCAG21/. Accessed 7 Aug 2025.
8. U.S. Census Bureau. Language spoken at home: 2022 American Community Survey 1-Year Estimates. 2023. https://data.census.gov/table?q=language+spoken+at+home&tid=ACSST1Y2022.S1601. Accessed 7 Aug 2025.
9. U.S. Department of Health & Human Services. Health Insurance Portability and Accountability Act of 1996 (HIPAA). 2023. https://www.hhs.gov/hipaa/for-professionals/security/guidance/index.html. Accessed 7 Aug 2025.
10. European Union. Regulation (EU) 2016/679 of the European Parliament and of the Council of 27 April 2016 (General Data Protection Regulation). 2016. https://eur-lex.europa.eu/eli/reg/2016/679/oj. Accessed 7 Aug 2025.
11. Zhang X, Saltman R. Impact of electronic health record interoperability on telehealth service outcomes. JMIR Med Inform. 2022;10(1):e31837. https://doi.org/10.2196/31837. PMID: 34890347; PMCID: PMC8790688.
12. Li E, Clarke J, Ashrafian H, Darzi A, Neves AL. The impact of electronic health record interoperability on safety and quality of care in high-income countries: systematic review. J Med Internet Res. 2022;24(9):e38144. https://doi.org/10.2196/38144. PMID: 36107486; PMCID: PMC9523524.
13. Dahlhausen F, Zinner M, Bieske L, Ehlers JP, Boehme P, Fehring L. There's an app for that, but nobody's using it: insights on improving patient access and adherence to digital therapeutics in Germany. Digi Health. 2022:8. https://doi.org/10.1177/20552076221104672.

Chapter 20
Virtual Events and Webinars as Marketing Tools

Introduction

Virtual events and webinars have emerged as powerful tools in the healthcare industry. They offer a dynamic way to engage both healthcare professionals and patients, allowing for the exchange of valuable information, updates on medical advancements, and discussions on critical health issues.

These digital platforms eliminate the geographical barriers of traditional in-person events. They provide a cost-effective and scalable format for outreach, professional development, and patient education. Healthcare organizations use virtual events to reach underserved communities, enhance communication, and promote health literacy.

The following sections explore how to plan, execute, and evaluate virtual events in healthcare. Topics include content creation, technical setup, audience engagement, legal considerations, and marketing impact. To begin, let's examine what makes virtual events effective in the healthcare context and how they differ from traditional outreach methods.

Why and How Virtual Events Work in Healthcare

Virtual events are online gatherings that enable participants to connect, learn, and interact in real-time from any location. They range from small-topic webinars to large-scale, multi-speaker healthcare conferences. These events are hosted on digital platforms that support presentations, Q&A sessions, breakout rooms, and live polling features.

In healthcare, producing a virtual event offers multiple benefits. Providers can reach broader and more diverse audiences, including patients in remote or

C. Quinn, *The Digital Practice*, https://doi.org/10.1007/978-3-032-07745-5_20

underserved communities. Virtual events are also cost-effective, eliminating travel and venue expenses while enabling on-demand participation.

These platforms encourage live feedback, strengthen engagement, and foster community among attendees. A hospital, for example, might host a webinar on managing heart disease that includes a Q&A session with a cardiologist, interactive polls, and follow-up materials.

Beyond education, virtual events support marketing goals by helping to showcase provider expertise, generate leads, expand brand visibility, and build trust. For practices and systems aiming to grow their digital footprint, webinars offer a strategic and measurable tool that enhances both outreach and patient relationships.

Once you understand the benefits, the next step is to create content that captures attention and encourages participation.

Content Creation for Virtual Events

Quality content keeps your audience engaged, builds trust, and enhances retention. It demonstrates that you understand their needs and encourages them to return for more. Creating compelling content for virtual events and webinars is crucial for engaging and educating your audience in the healthcare sector. This section provides guidance on crafting content that resonates with participants, whether they are healthcare professionals or patients.

Addressing Patient Concerns Start with content that directly addresses common patient concerns. This could include discussions on managing chronic conditions, navigating healthcare services, or understanding treatment options. For example, a webinar could feature a Q&A session with healthcare professionals discussing strategies for managing diabetes or hypertension, providing valuable insights and practical advice for patients.[1]

Showcasing Medical Advancements Highlighting the latest advancements in medical technology and treatments can captivate your audience. Consider featuring expert talks on new diagnostic tools, innovative therapies, or breakthrough research findings. For instance, a virtual event could showcase the latest in minimally invasive surgical techniques, offering attendees a glimpse into the future of healthcare.[2]

Providing Health and Wellness Tips Content that offers health and wellness tips can be highly engaging and beneficial for attendees. This might include sessions on nutrition, exercise, mental health, or preventive care. For example, a series of webinars could focus on "30 Days of Wellness," offering daily tips for improving physical

[1] https://www.centrum.com/learn/articles/18-wellness-tips-for-a-healthier-you/. Accessed 3/28/24.

[2] https://www.mountsinai.org/about/newsroom/2024/mount-sinai-leaders-showcase-innovations-and-solutions-in-health-delivery-and-care-quality-at-arab-health-2024. Accessed 3/28/24.

and mental health, such as staying hydrated, incorporating physical activity into daily routines, and practicing mindfulness.[3]

By incorporating these types of content into your virtual events and webinars, you can create engaging and informative presentations that address the needs and interests of your audience. Remember, the goal is to provide value, whether by solving problems, offering new information, or inspiring healthier lifestyles.

Strong content not only serves the audience, but it also positions your organization as a leader. This creates significant marketing opportunities.

Figure 20.1 highlights three questions to ask before you publish content.

Effective Marketing Tools Virtual events and webinars serve as effective marketing tools by showcasing the expertise of healthcare professionals, highlighting the services and technologies of healthcare organizations, and building trust with patients and the community. They offer an interactive platform for engaging with current and potential patients, educating them on important health topics, and promoting health literacy. By leveraging virtual events, healthcare marketers can effectively communicate their message, expand their reach, and ultimately improve patient care and outcomes.

Virtual events and webinars represent a significant shift in how healthcare information is shared and consumed. They not only offer practical solutions to the challenges of traditional event hosting but also open new avenues for marketing and patient engagement in the healthcare industry.[4,5]

Fig. 20.1 Three questions to ask before publishing content. (Image Quinn/ChatGPT)

QUICK TIP BOX:
3 QUESTIONS TO ASK BEFORE PUBLISHING CONTENT

1. Does this content address my audience's current needs or interests?
2. Will it encourage engagement, sharing, or action?
3. Is the tone and language consistent with my brand voice?

[3] https://www.unclenoir.org/wellness-connection-blog/exercise-nutrition/30-days-of-wellness/. Accessed 3/28/24.

[4] https://www.beaconlive.com/blog/how-the-healthcare-industry-is-using-webinars. Accessed 3/28/24.

[5] https://www.healthcarestrategy.com/blog/4-reasons-healthcare-marketers-embrace-virtual-events/. Accessed 3/28/24.

But to make the most of these opportunities, you need the right tools and infrastructure. That starts with your technical setup. Table 20.1 lists common marketing goals and how virtual events support each one.

Technical Setup for Hosting Virtual Events For healthcare providers venturing into digital marketing through virtual events and webinars, understanding the technical setup is essential. This section will guide you through selecting the right platforms, equipment, and software to ensure a successful virtual event.

Choosing the Right Platform Your choice of platform is foundational to the success of your virtual event. Platforms like Zoom offer robust features for hosting webinars, including live streaming options, breakout rooms, and interactive Q&A sessions. It's important to define and visualize your event experience early on, ensuring the platform you choose can support your vision.[6]

Essential Equipment The quality of your presentation can be significantly impacted by the equipment you use. At a minimum, you will need a reliable computer, a high-quality webcam, and a professional microphone. This setup ensures clear visuals and audio, which are critical for maintaining audience engagement. Investing in good lighting can also enhance the video quality.[7]

Software Needs Beyond the platform, specific software may be required for presentation delivery, audience interaction, and event analytics. Presentation software like PowerPoint or Keynote is essential for creating engaging slides. Additionally, audience interaction tools such as Slido or Mentimeter can facilitate live polls and Q&A sessions, making the webinar more interactive and engaging.[7]

Table 20.1 Marketing goals achieved by virtual events

Goal	How virtual events support it
Brand visibility	Positions providers as experts and increases exposure to new audiences
Patient education	Delivers trusted, interactive information on conditions, treatments, and prevention
Lead generation	Captures contact information through registration, polls, and Q&A participation
Trust building	Showcases clinical knowledge in a transparent, real-time format
Content repurposing	Generates on-demand videos, blog topics, and social media posts from one event
Community engagement	Fosters dialogue, answers questions, and creates recurring touchpoints with patients and the public

[6] https://www.zoom.com/en/products/event-platform/resources/virtual-events-guide/. Accessed 3/28/24.

[7] https://www.actualtechmedia.com/blog/how-to-host-a-tech-webinar-the-ultimate-guide/. Accessed 3/28/24.

By carefully selecting your platform, ensuring you have the right equipment, and utilizing effective software, you can create a professional and engaging virtual event. This technical foundation not only enhances the experience for participants but also positions your healthcare practice as a forward-thinking and patient-centric organization.[8]

High-quality production helps position your practice as professional and trustworthy. Figure 20.2 provides a basic tech setup checklist for healthcare webinars.

Engaging Your Audience Engaging your audience throughout the lifecycle of a virtual event or webinar is essential for maximizing participation and ensuring a memorable experience. Here are strategies to engage your audience before, during, and after your event:

Fig. 20.2 Top five mistakes that reduce webinar engagement. (Image Quinn/ChatGPT)

[8] https://www.banzai.io/blog/webinar-marketing-strategy-guide. Accessed 3/28/24.

Before the Event:

Promotion and Teasers: Utilize social media, email newsletters, and your website to promote your event. Share teasers of the content or speakers to build anticipation.[9]

Interactive Registration: Enhance the registration process by asking attendees to submit questions they hope to have answered. This not only engages them early on but also helps tailor the content to the audience's specific interests.

During the Event:

Live Polls: Incorporate live polls during your presentation to gather real-time feedback and opinions from your audience. This keeps participants active and involved.[10]

Q&A Sessions: Allocate time for a Q&A session where attendees can ask questions directly to the speakers, fostering a sense of community and interaction.[11]

Breakout Rooms: For larger events, use breakout rooms to divide the audience into smaller groups for more focused discussions. This can enhance engagement by allowing more personalized interaction.[11]

After the Event:

Follow-up Surveys: Send out surveys to gather feedback on what attendees liked and what could be improved. This shows that you value their opinion and are committed to enhancing future events.[9]

Content Sharing: Share recordings of the event, along with additional resources mentioned during the webinar.[12] This keeps the conversation going and provides value even after the event has ended.

Feedback Surveys: Send out post-event surveys to collect attendees' feedback. This is vital for understanding what worked well and what could be improved. Honest feedback helps in tailoring future events to better meet the needs of your audience.

Continuous Engagement: Establish a schedule for regular communication with your audience through newsletters, social media updates, and announcements of upcoming events. This helps in building a community around your brand and keeps your audience engaged over the long term.[13]

Impact Assessment: Analyze the feedback and engagement metrics to assess the event's impact on patient care and practice growth. This could involve

[9] https://www.hamilton.edu/academics/centers/oralcommunication/guides/how-to-engage-your-audience-and-keep-them-with-you. Accessed 3/28/24.

[10] https://www.demio.com/blog/webinars-vs-virtual-events. Accessed 3/28/24.

[11] https://www.experiencewelcome.com/blog/virtual-event-vs-webinar-what-is-the-difference. Accessed 3/28/24.

[12] https://spotme.com/blog/post-event-engagement/. Accessed 3/28/24.

[13] https://splashthat.com/resources/event-marketing-guide/post-event-engagement-strategies. Accessed 3/28/24.

tracking the number of follow-up appointments made, the increase in patient inquiries, or the growth in social media followers as a result of the event [1].

Community Building: Encourage attendees to join your online community, whether it's a social media group or a forum on your website. This keeps them engaged with your brand and fosters long-term relationships.

By implementing these strategies, healthcare providers can create virtual events and webinars that not only educate but also actively engage their audience, leading to higher satisfaction and better outcomes.

As engagement increases, so does responsibility. The next section outlines how to stay legally compliant and ethically sound. Table 20.2 provides a basic tech setup checklist for healthcare webinars.

Legal and Ethical Considerations When using virtual healthcare events as digital marketing, legal and ethical considerations are paramount, especially concerning patient privacy and data security. This section delves into the critical aspects healthcare providers must consider to ensure compliance with healthcare regulations, notably the Health Insurance Portability and Accountability Act (HIPAA) in the United States.

1. *HIPAA Compliance*: HIPAA sets the standard for protecting sensitive patient data.[14] Any healthcare provider that deals with protected health information (PHI) must ensure that all the required physical, network, and process security measures are in place and followed. For virtual events, this means ensuring that any patient information shared or discussed is secured against unauthorized access.

Table 20.2 Basic tech setup for healthcare webinars: checklist

Item	Description
Reliable internet	High-speed connection (wired preferred) with at least 10 Mbps upload speed
Webcam	HD webcam positioned at eye level for clear visuals
Microphone	External USB microphone or headset for crisp, professional audio
Lighting	Front-facing light source or ring light to eliminate shadows
Background	Neutral, uncluttered background or virtual backdrop branded to your practice
Platform	HIPAA-compliant webinar platform (e.g., Zoom for Healthcare, Webex)
Slide deck	Professionally designed slides in widescreen format (16:9)
Backup plan	Plan for technical issues (backup device, tech contact on standby)
Recording tools	Enable recording settings for post-event use and compliance archiving
Test run	Run a full technical rehearsal at least 24 hours before the event

[14] https://www.hhs.gov/hipaa/for-professionals/security/laws-regulations/index.html. Accessed 3/28/24.

2. *Informed Consent*: It's crucial to obtain informed consent from participants before collecting, using, or sharing their health information. This includes informing them about the risks and benefits of remote therapy or consultation. Clear communication about how their data will be used and protected is essential [2].
3. *Maintaining Patient Privacy*: Virtual event platforms must be secure and comply with privacy laws to protect patient confidentiality. This involves using encrypted communication channels and secure data storage solutions to prevent data breaches [3].
4. *Ethical Research and Privacy Protections*: Ethical health research and privacy protections both offer significant benefits to society. When conducting research or sharing case studies during virtual events, it's vital to anonymize patient data to prevent the identification of individuals from shared information [4].
5. *Telemedicine Platforms Security*: Telemedicine platforms used for virtual healthcare events should have robust security measures in place. This includes end-to-end encryption for video calls, secure messaging, and adherence to national and international data protection regulations.[15]

By adhering to these legal and ethical considerations, healthcare providers can ensure that their virtual events not only comply with regulations but also foster trust with their audience. This trust is crucial for the successful adoption and effectiveness of virtual healthcare services.

With strategy, technology, engagement, and compliance aligned, healthcare providers are ready to take action.

Conclusion

Virtual events and webinars have become essential components of modern healthcare marketing. They offer powerful ways to reach large and geographically diverse audiences while delivering timely, credible information. These digital formats promote two-way interaction, support patient education, and help build trust in your organization. When executed well, they not only raise awareness but also contribute to better health outcomes.

To get started, identify a key health topic, choose a reliable webinar platform, and promote the event through email campaigns and social media channels.

[15] https://mindspaceconferences.com/digitalhealth/sessions/ethical-and-legal-considerations-in-telemedicine/. Accessed 3/28/24.

References

1. Purcarea VL. The impact of marketing strategies in healthcare systems. J Med Life. 2019;12(2):93–6. https://doi.org/10.25122/jml-2019-1003. PMID: 31406509; PMCID: PMC6685306.
2. Solimini R, Busardò FP, Gibelli F, Sirignano A, Ricci G. Ethical and legal challenges of telemedicine in the era of the COVID-19 pandemic. Medicina (Kaunas). 2021;57(12):1314. https://doi.org/10.3390/medicina57121314. PMID: 34946259; PMCID: PMC8705012.
3. Fields BG. Regulatory, legal, and ethical considerations of telemedicine. Sleep Med Clin. 2020;15(3):409–16. https://doi.org/10.1016/j.jsmc.2020.06.004. Epub 2020 July 7. PMID: 32762973; PMCID: PMC7340020. Accessed 28 Mar 2024.
4. Institute of Medicine (US) Committee on Health Research and the Privacy of Health Information: The HIPAA Privacy Rule. In: Nass SJ, Levit LA, Gostin LO, editors. Beyond the HIPAA privacy rule: enhancing privacy, improving health through research. Washington, DC: National Academies Press; 2009. 2, The value and importance of health information privacy. Available from: https://www.ncbi.nlm.nih.gov/books/NBK9579/.

Chapter 21
Crisis Management and Communication in Digital Healthcare Marketing

In digital healthcare marketing, a crisis can be defined as any adverse event, review, or issue that gains public traction and has the potential to harm the reputation or operational capacity of a healthcare organization.[1] These crises often arise from the rapid spread of information online, where a single incident can quickly escalate if not managed properly. The unique challenges of healthcare marketing, including the handling of sensitive patient information and compliance with healthcare regulations, make effective crisis management crucial.[2] Managing these events swiftly and strategically can make the difference between a temporary disruption and lasting damage to an organization's credibility.

Examples of Real-World Crises in Digital Healthcare Marketing

In the digital era, healthcare organizations face an expanding range of reputational and operational threats. From data breaches to misinformation campaigns, each crisis type presents unique challenges that demand tailored responses. This section outlines key categories of digital healthcare crises and provides real-world examples to illustrate how these incidents unfold and how organizations can respond effectively. Understanding these patterns prepares marketing teams to detect early warning signs, coordinate responses, and maintain public trust under pressure. The following case types represent the most common and high-impact crises that digital healthcare teams encounter.

[1] https://wearecsg.com/blog/pr-healthcare-crisis/. Accessed 8/7/25.

[2] https://www.weareamnet.com/blog/healthcare-marketing-challenges/. Accessed 8/7/25.

C. Quinn, *The Digital Practice*, https://doi.org/10.1007/978-3-032-07745-5_21

Data Breaches

Data breaches remain among the most frequent and damaging digital crises in healthcare. These events not only compromise sensitive patient information but also erode public trust, trigger legal and regulatory responses, and cause significant financial harm.

A major example is the 2015 cyberattack on Anthem Inc., in which hackers gained unauthorized access to the names, birthdates, medical IDs, Social Security numbers, and employment information of nearly 79 million individuals. The breach led to widespread public outcry, a $115 million class-action settlement, and intense regulatory scrutiny from the Department of Health and Human Services (HHS) and state attorneys general [1].

Unfortunately, large-scale breaches are not rare. In 2023 alone, over 133 million healthcare records were exposed due to hacking incidents in the United States, according to the HHS Office for Civil Rights breach portal.[3] These breaches are increasingly driven by ransomware attacks that paralyze hospital systems, delay care, and force diversion of emergency services.

The consequences go beyond immediate financial losses. A 2021 study found that healthcare data breaches were associated with longer hospital stays,[4] increased mortality,[5] and delayed treatment in the weeks following the incident [2], highlighting the real-world impact on patient care.

To prevent and contain breaches, healthcare organizations must:

- Conduct regular cybersecurity audits.
- Implement endpoint detection and response (EDR) tools.
- Train staff to recognize phishing and social engineering tactics.
- Create and routinely test incident response and recovery protocols.

The reputational damage from a breach is amplified when organizations delay disclosure or appear unprepared. Transparent, timely communication with patients and regulators, along with visible remediation steps, is critical to maintaining trust during and after a breach.

While security breaches often involve private data, misinformation attacks target public perception and can spread just as rapidly.

---

[3] Civil Rights Breach Portal https://ocrportal.hhs.gov/ocr/breach/breach_report.jsf. Accessed 8/5/25.

[4] https://www.censinet.com/blog/new-ponemon-institute-research-shows-ransomware-attacks-on-healthcare-delivery-organizations-can-lead-to-increased-mortality-rate?utm_source=chatgpt.com. Accessed 8/5/25.

[5] The HIPAA Journal. https://www.hipaajournal.com/study-confirms-increase-in-mortality-rate-and-poorer-patient-outcomes-after-cyberattacks/?utm_source=chatgpt.com. Accessed 8/2/25.

Misinformation Spread

False or misleading health claims spread rapidly online, especially when information comes from unofficial or automated sources. During the COVID-19 pandemic, misleading content about the virus, vaccines, and treatments circulated widely across social media platforms. Hospitals and public health agencies had to respond quickly using accurate, accessible content. Many developed rapid-response teams to counter myths in real time. Platforms like Facebook, Instagram, and YouTube were used to disseminate science-based messages, reduce panic, and redirect patients toward credible resources.

Beyond COVID-19, misinformation has evolved. In 2023, researchers identified a growing trend in AI-generated health advice that appeared convincing but was medically incorrect. One widely circulated example involved a chatbot embedded in a weight-loss app that provided dangerous advice about fasting and caloric intake. The chatbot recommended extreme restrictions without screening for patient age, health conditions, or eating disorder risk factors. After public backlash, the company paused the tool and issued a statement clarifying the limitations of its AI engine.

Health systems must now anticipate the spread of false content created not only by individuals but also by generative AI. This includes deepfake physician endorsements, fake diagnostic tools, and symptom-checkers producing hallucinated recommendations. Monitoring trending keywords and misinformation vectors, such as health subreddits, TikTok health challenges, and GPT-generated posts, should become a core function of digital marketing and public communication teams.

Proactive mitigation involves creating ready-to-publish fact sheets, short-form video responses, and using verified channels to flag or correct false claims. Collaborating with platform moderators or using automated alert systems can also help flag trending misinformation before it gains traction.

While misinformation poses an external communication challenge, operational failures can just as easily erode trust when digital systems break down during patient care.

Service Failures

Service failures in digital healthcare are not just technical mishaps; they are perceived as breaches of trust. When patients receive incorrect lab results, experience canceled procedures due to electronic health record (EHR) outages, or face poor communication during virtual visits, the impact extends beyond the immediate clinical consequences. These experiences undermine patient confidence, fuel dissatisfaction, and can rapidly evolve into reputational crises, especially when shared on social media or review sites.

To manage these risks, healthcare organizations must implement structured response plans. Transparency is critical. Patients should receive clear explanations of what happened, how the issue will be resolved, and what safeguards are being put in place to prevent recurrence. Offering proactive remediation, such as follow-up appointments, personal calls, or service recovery credits, can help rebuild trust and prevent viral fallout.

Healthcare systems should also monitor sentiment on digital platforms and set up alerts for keywords indicating service disruptions. A swift, empathetic, and coordinated response across clinical and communication teams can help contain reputational damage and demonstrate accountability.

Callout Box: Avoid These Digital Crisis Pitfalls

Avoid These Digital Crisis Pitfalls

Digital missteps in healthcare communication are often preventable—but costly. These common pitfalls can quickly escalate a manageable situation into a reputational or regulatory crisis:

- **Delaying breach disclosure**
  Breach notification laws require a timely response. Delays risk fines, patient backlash, and media scrutiny. Prepare templated alerts and escalation pathways before an incident occurs.
- **Ignoring online reviews and feedback during outages**
  Patients post real-time complaints on Google, Yelp, and social media. Failure to monitor or respond gives the impression of disorganization. Use sentiment monitoring tools to catch early warning signs.
- **Using vague or generic language in serious situations**
  Patients want clarity, not corporate jargon. Messaging should directly address the nature of the incident, what is affected, and what's being done to fix it.
- **Over-relying on legal boilerplate in public updates**
  Overly cautious, legalistic statements often appear evasive. Work with legal and communications teams to ensure responses are both compliant and human.

Beyond communication missteps, healthcare organizations also face reputational risks related to financial misconduct and billing integrity.

Financial Fraud

Instances of billing fraud, unauthorized charges, or insurance abuse can provoke both legal action and public backlash [3]. For example, whistleblower claims or news stories about inflated charges can go viral, casting doubt on an organization's ethics. Digital teams must monitor financial complaints on review platforms and respond promptly with factual, legal-safe messaging.

While financial fraud often comes from external manipulation, insider threats originate from within the organization and require a different approach.

Insider Threats

Employees or authorized insiders sometimes cause data breaches through careless, negligent, or intentional misuse of PHI. These breaches are a leading cause of healthcare data incidents and can involve sharing or selling patient information, unauthorized access, or disclosure through digital platforms, including social media. Even accidental leaks are considered HIPAA violations and require investigation and remediation.[6] Even one staff member acting inappropriately online can draw media attention and regulatory fines. Regulatory agencies, such as the U.S. Department of Health and Human Services' Office for Civil Rights (OCR), impose strict penalties for unauthorized disclosure of PHI, including fines up to $250,000 and prison sentences for intentional or egregious violations. Penalties are often more severe when the breach garners public attention, damaging both trust and reputation.[7] A crisis plan must include internal communication policies and clear escalation paths for internal breaches.

Managing insider incidents requires a nuanced understanding of HIPAA obligations and what can be safely communicated to the public during an investigation.

Sidebar: HIPAA in a Crisis—What You Can and Cannot Say

During a breach or insider incident, public communication must follow HIPAA's Breach Notification Rule (45 CFR §§ 164.400–414). Avoiding violations while maintaining transparency requires careful attention to content, timing, and tone.

What must be reported to HHS:

- Any breach of unsecured protected health information (PHI).
- For breaches affecting 500 or more individuals, you must report to the U.S. Department of Health & Human Services (HHS) within 60 calendar days of discovery.
- For breaches affecting fewer than 500, report via the HHS breach portal within the calendar year.

(continued)

[6] The HIPAA Journal. https://www.hipaajournal.com/insider-threats-to-healthcare-records/. Accessed 8/2/25.

[7] Accountable. https://www.accountablehq.com/post/consequences-of-not-following-hipaa-laws. Accessed 8/4/25.

When and how to notify affected individuals:

- Notify patients without unreasonable delay, and no later than 60 calendar days after discovering the breach.
- Notices must include:
  - What happened and when.
  - What information was involved.
  - Steps the organization is taking.
  - How individuals can protect themselves.
  - Contact information for questions.
- Use first-class mail or secure email if the patient has agreed to electronic notices.

When media notification is required:

- If a breach affects 500 or more people in a single state or jurisdiction, you must notify prominent media outlets serving that area.

What not to say:

- Do not confirm patient names or health information in public statements.
- Avoid implying diagnoses, treatments, or individual fault.
- Never post screenshots or internal emails, even if redacted.

HIPAA-safe public statement example:

We identified unauthorized access to patient information on [date]. We immediately secured the system, launched an investigation, and notified all affected individuals. We have reported the incident to federal regulators and are offering identity protection services.

It is very important to retain documentation of all communications for regulatory audits.

When in doubt, coordinate with legal and compliance teams before releasing any public messaging.

Real-world failures involving operations, reputation, or regulatory breaches illustrate the importance of proactive planning and coordinated team responses.

Platform-Specific Failures

Failures in digital health platforms, such as telehealth systems, patient portals, or online appointment scheduling tools, undermine access to care and disrupt the patient experience. These outages are more than technical errors; they can cause missed appointments, treatment delays, and loss of trust, especially when patients feel uninformed or ignored during the disruption.

A documented case in 2023 involved Oracle engineers accidentally deleting data during routine maintenance, resulting in a five-day outage across 45 hospitals. The incident forced providers to revert to paper records, delayed care coordination, and led to widespread operational strain [4].

During the COVID-19 pandemic, surging demand for telehealth overwhelmed many platforms, leading to appointment delays and access issues. Some hospitals struggled to scale infrastructure quickly enough to meet patient needs, as reported by multiple digital health providers and healthcare systems [5].

Patients' experiences during these breakdowns shape their perception of a health system's competence and reliability. A 2024 study in *JMIR Aging* found that older adults using a digital communication platform to reach their general practitioners appreciated the convenience but struggled with usability barriers, including login issues and confusing automated workflows [6].

To mitigate fallout from these platform-specific failures, healthcare organizations should implement a response plan that includes:

- Immediate outage alerts via SMS, email, and portal banners.
- A centralized system status page to ensure transparency.
- Active monitoring of social media for complaints or misinformation.
- Clear post-resolution messaging and rescheduling procedures.

A fast, coordinated communication strategy helps reduce patient frustration and protects the organization's reputation during digital service failures.

Crises in digital healthcare marketing come in many forms, from security breaches to operational breakdowns. Each requires a tailored response strategy grounded in transparency, speed, and digital fluency. Anticipating these risks and practicing scenario-based planning helps safeguard both reputation and patient trust. See "Appendix 21.1: Digital Crisis Simulation Worksheet" for a practical tool to guide internal crisis exercises or real-time incident management.

Given the wide range of digital threats healthcare organizations face, effective preparation becomes essential, not only to minimize harm but to ensure a rapid, coordinated response when a crisis emerges.

Whether triggered by technical disruption or human error, each of these incidents underscores the need for strong planning and digital responsiveness.

Preparing for Crises in the Digital Age

Preparation is key to effective crisis management. However, preparation involves more than just having a plan; it requires understanding the unique challenges of healthcare marketing, such as regulatory constraints and the sensitivity of health-related information.[8]

[8] https://www.o8.agency/blog/6-healthcare-marketing-challenges-and-how-solve-them. Accessed 3/28/24.

Preparation involves establishing a crisis communication plan that outlines specific steps to take when a crisis emerges. The plan should identify key team members responsible for crisis response, define potential crisis scenarios, and establish communication protocols. Utilizing digital tools for real-time monitoring of online conversations about your brand can also provide early warnings of potential crises.[9] Figure 21.1 outlines crisis planning steps from team roles to monitoring and mapping responsibilities before incidents.

Fig. 21.1 Crisis planning. (Image Quinn/ChatGPT)

[9] https://www.linkedin.com/pulse/effective-crisis-management-digital-age-zainullah-khan-9josf/?trk=public_post_main-feed-card_feed-article-content. Accessed 3/28/24.

Best Practices for Transparent and Effective Communication During a Crisis

In healthcare, crisis communication is not simply about issuing a press release or social media post. It's about preserving trust when systems fail, expectations are unmet, or lives are disrupted. The stakes are high. Poor communication during a crisis can lead to reputational damage, regulatory scrutiny, or long-term erosion of patient confidence. The following best practices offer a framework for healthcare leaders and marketers to respond effectively using real-world strategies and confirmed examples. Figure 21.2 shows a concise workflow from detection to message drafting, review, approval, and release across channels.

1. **Act Quickly**

The earliest phase of a crisis is often called the prodromal phase. This is when warning signs emerge, but the full impact hasn't yet unfolded. Organizations that act decisively during this window can shape the narrative and reduce speculation. Waiting too long allows misinformation to spread unchecked.

During the early weeks of the COVID-19 pandemic, many healthcare systems leveraged digital platforms like Twitter and Facebook to issue rapid updates about clinic closures, telehealth availability, and infection control policies [7]. These

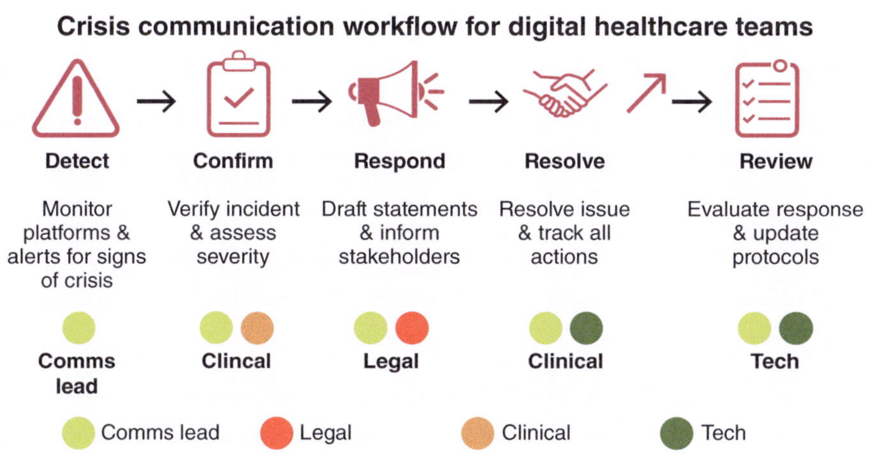

Fig. 21.2 Crisis communication workflow. (Image Quinn/ChatGPT)

real-time updates helped keep patients informed, reduced call center overload, and demonstrated proactive leadership during a time of global uncertainty.[10]

2. **Be Transparent**

Transparency is not a courtesy; it's a risk management tool. During a crisis, patients and the public need direct, unambiguous information about what happened, how it affects them, and what the organization is doing in response. Evasion or minimization undermines credibility.

The Cleveland Clinic's response to a 2023 data breach illustrates this well. Once unauthorized access to a staff email account was discovered, the clinic quickly notified approximately 83,000 affected patients.[11] Notifications included a summary of the breach, information about what data was exposed, and steps the clinic was taking to improve security. This timely, detailed disclosure helped contain reputational fallout and reassured patients that their data was being handled responsibly.[12]

3. **Use Multiple Channels**

No single communication method reaches everyone. During a crisis, organizations must communicate across multiple platforms to account for different access needs, digital literacy levels, and urgency.

The Centers for Disease Control and Prevention (CDC) modeled this during the 2009 H1N1 influenza outbreak. It used a combination of social media, website updates, email newsletters, and press briefings to distribute consistent messages. The CDC's approach ensured that both healthcare providers and the general public received timely updates through their preferred channels, minimizing confusion and promoting alignment across diverse audiences.[13]

4. **Monitor and Adjust**

Crises unfold quickly. Public sentiment can shift in hours alongside false narratives and speculation. Timely monitoring enables healthcare organizations to detect emerging concerns, guide corrective messaging, and maintain public trust.

Example In early 2025, a ransomware attack targeted DaVita, a U.S. provider of kidney care services.[14] Cybercriminals accessed sensitive medical and personal records of nearly one million patients, including names, Social Security numbers,

---

[10] https://www.statnews.com/2020/03/17/telehealth-services-overwhelmed-amid-coronavirus-pandemic/?utm_source=chatgpt.com. Accessed 8/8/25.

[11] https://www.mass.gov/doc/assigned-data-breach-number-29674-medinform-inc/download. Accessed 8/10/25.

[12] https://www.jdsupra.com/legalnews/medinform-inc-announces-data-breach-6424529/. Accessed 8/8/25.

[13] https://emergency.cdc.gov/h1n1flu/communication/pdf/h1n1-comms-evaluation.pdf. Accessed 8/7/25.

[14] The HIPAA Journal. https://www.hipaajournal.com/davita-ransomware-attack/#:~:text=DaVita%20described%20the%20cyberattack%20as,to%20individual%20and%20may%20include. Accessed 8/7/25.

and health insurance details. DaVita responded by notifying law enforcement, alerting patients via letters, and offering identity protection services.[15] The company also provided guidance on monitoring accounts and credit for signs of fraud. This multichannel response helped address immediate public concern and provided actionable reassurance to affected individuals.

5. **Conduct Post-Crisis Evaluation**

Post-crisis reviews are often neglected once the incident has passed. However, the most resilient healthcare organizations treat each crisis as an opportunity to learn and improve. Leadership should pause to evaluate what happened, who was affected, and how future responses can improve.

Following the 2014 Ebola outbreak, Doctors Without Borders (MSF) conducted an internal and external evaluation that revealed gaps in communication. Teams in the field struggled with unclear messages, inconsistent information flow between headquarters and frontline workers, and poor coordination with external agencies [8]. Some messages failed to reach their intended audiences, while others caused confusion due to a lack of targeted messaging.

To address these issues, MSF issued several reforms [9]:

- Created defined roles for health promotion and communication staff.
- Developed clearer strategies for advocacy and messaging.
- Deployed local communication tools that could adapt to changing field conditions.
- Added full-time communications staff at both headquarters and field offices.
- Documented effective practices for future reference.
- Prioritized training and message development protocols.

The MSF review also noted a lack of time for strategic reflection, which added to communication errors during the crisis.

Healthcare systems and digital marketing teams can learn from this approach. They should conduct structured reviews that examine both internal coordination and public messaging [10]. These reviews should go beyond technical performance and include legal, ethical, and reputational concerns [7].

Key questions to ask in a post-crisis review:

- Did staff follow the approved crisis communication plan?
- Were social media responses consistent with HIPAA and other legal policies?
- Were messages timely, accurate, and easy to understand?
- Did patients or the public misunderstand key messages?
- Did leadership escalate issues promptly and through the correct channels?
- Were any system vulnerabilities exploited that must be addressed?

---

[15] https://www.tomsguide.com/computing/online-security/over-900-000-hit-in-massive-healthcare-data-breach-names-addresses-and-social-security-numbers-exposed-online?utm_source=chatgpt.com. Accessed 8/7/25.

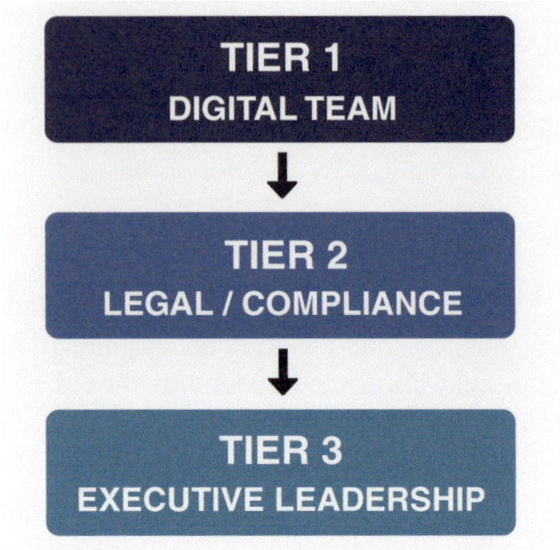

Fig. 21.3 Crisis escalation pathway in healthcare marketing response. (Image Quinn/ChatGPT)

A formal post-crisis evaluation identifies specific gaps, provides a foundation for training, and builds institutional memory. It ensures that your next response is faster, clearer, and safer for both your organization and the patients you serve.

To ensure clarity and accountability during crisis response, teams must follow a defined escalation path. Figure 21.3 presents a tiered escalation path for marketing response, from frontline handling to executive leadership.

Conclusion

Healthcare organizations cannot prevent every crisis, but they can control how they respond. Acting quickly, communicating transparently, using multiple channels, monitoring public sentiment, and conducting rigorous after-action reviews form the backbone of an effective communication strategy. In an era of digital immediacy, well-executed crisis communication is not just protective—it's strategic.

Appendix 21.1: Digital Crisis Simulation Worksheet

This worksheet helps healthcare organizations simulate or respond to digital crises in real time. It guides planning, coordination, communication, and evaluation, key to minimizing harm and maintaining public trust.

Crisis Profile

Field	Response
Simulation or Real Event?	☐ Simulation ☐ Actual Event
Type of Crisis	☐ Data Breach ☐ Misinformation ☐ Service Outage ☐ Insider Threat ☐ Financial Fraud ☐ Other: ______
Date Discovered	________________________________
Affected Systems or Services	________________________________
Estimated Affected Individuals	________________________________
Known Risks	☐ HIPAA exposure ☐ Operational downtime ☐ Reputation loss ☐ Regulatory reporting ☐ Patient safety ☐ Legal action

---

Core Response Team

Role	Assigned Staff	Contact Info
Incident Commander	____________________ —	____________________ —
Communications Lead	____________________ —	____________________ —

Role	Assigned Staff	Contact Info
Legal/Compliance Officer	____________________ —	____________________ —
Privacy Officer (HIPAA)	____________________ —	____________________ —
Clinical Liaison	____________________ —	____________________ —
IT/Cybersecurity Lead	____________________ —	____________________ —
Spokesperson (if different)	____________________ —	____________________ —

Initial Messaging Strategy

Current Facts (What can be confirmed?)

____

____

Initial Public Statement (Holding Statement):

"We are investigating a disruption involving part of our systems. At this time, there is no confirmed evidence of patient harm. We are working with internal teams and external experts to understand the scope and will share updates as appropriate."

Communication Channels to Use:

☐ Website Banner

☐ Email to Affected Users

☐ SMS Alert

☐ Patient Portal Notification

☐ Social Media (Specify Platforms: _______)

☐ News Media

☐ Direct Provider Messaging

Regulatory & Stakeholder Notification Checklist

Stakeholder	Required	Method of Contact	Completed
Patients	☐		☐
HHS (if PHI breach)	☐		☐
Media (if >500 in jurisdiction)	☐		☐
Internal Staff	☐		☐
Partner Organizations	☐		☐
Board / Executive Leadership	☐		☐

Crisis Timeline Snapshot

Action	Time Completed
Crisis identified	____________________ —
Core team activated	____________________ —
First internal update	____________________ —
First external communication sent	____________________ —
Regulators notified (if applicable)	____________________ —

Post-Crisis Review

What worked well during the response?

What caused delays or confusion?

Were communications clear and appropriate?

☐ Yes

☐ No → Comments: ______________________

Follow-Up Actions:

☐ Update incident response policy

☐ Conduct after-action debrief

☐ Improve cross-department coordination

☐ Revise message templates

☐ Update training based on lessons learned

Prepared by: ______________________

Date Completed: ____________________

Approved by: ______________________

References

1. U.S. Department of Health & Human Services. Anthem pays OCR $16 million in record HIPAA settlement following largest U.S. health data breach. 2018, October 15. https://www.hhs.gov/hipaa/for-professionals/compliance-enforcement/agreements/anthem/index.html. Accessed 7 Aug 2025.
2. Choi SJ, Johnson ME, Lehmann CU. Data breach remediation efforts and their implications for hospital quality. Health Serv Res. 2019;54(5):971–80. https://doi.org/10.1111/1475-6773.13203. PMID: 31506956; PMCID: PMC6736905.
3. U.S. Department of Justice. Justice Department recovers over $2.2 billion from False Claims Act cases in fiscal year 2020. 2021. https://www.justice.gov/opa/pr/justice-department-recovers-over-22-billion-false-claims-act-cases-fiscal-year-2020
4. Economic Times. Ctrl + Alt + disaster: how oracle techies' wrong click brought nearly 50 hospitals to its knees for 5 days. 2024, February 11. https://m.economictimes.com/news/

international/global-trends/ctrlaltdisaster-how-oracle-techies-wrong-click-brought-nearly-50-hospitals-to-its-knees-for-5-days/articleshow/120753264.cms. Accessed 8 Aug 2025.
5. Ross C. Telehealth services overwhelmed amid coronavirus pandemic. Stat News. 2020, March 17. https://www.statnews.com/2020/03/17/telehealth-services-overwhelmed-amid-coronavirus-pandemic/. Accessed 7 Aug 2025.
6. Knotnerus HR, Ngo HTN, Maarsingh OR, van Vugt VA. Understanding older adults' experiences with a digital health platform in general practice: qualitative interview study. JMIR Aging. 2024;7:e59168m. https://doi.org/10.2196/59168. PMCID: 11378695.
7. Shi W, Yeung T, Dmello J. Organizational use of social media during a global health crisis. Front Commun. 2024;9:Article 1421165. https://doi.org/10.3389/fcomm.2024.1421165. Accessed 8 Aug 2025.
8. National Academies of Sciences, Engineering, and Medicine. The Ebola epidemic in West Africa: proceedings of a workshop. National Academies Press; 2016. https://doi.org/10.17226/23653.
9. Toppenberg-Pejcic D, Noyes J, Allen T, Alexander N, Vanderford M, Gamhewage G. Emergency risk communication: lessons learned from a rapid review of recent gray literature on Ebola, Zika, and Yellow Fever. Health Commun. 2019;34(4):437–55. https://doi.org/10.1080/10410236.2017.1405488. Epub 2018 Mar 20.
10. Al Qaf'an E, Alford S, Porteous K, Lim D. Healthcare decision-making in a crisis: a qualitative systemic review protocol. Emerg Med Int. 2024;2024:2038608. https://doi.org/10.1155/2024/2038608. PMID: 38715979; PMCID: PMC11074907.

Chapter 22
Conclusion

This book has examined the full spectrum of digital marketing strategies tailored for healthcare providers. It outlined how to build a strong digital presence, increase patient engagement, and comply with industry regulations. Key chapters explored local SEO, digital content strategies, social media, mobile platforms, and HIPAA-compliant advertising. Additional sections addressed online reputation management, telehealth marketing, and the use of emerging technologies.

Marketing in healthcare is no longer a secondary concern. It plays a central role in patient acquisition, retention, and trust. An ethical, patient-centered digital strategy can expand your practice's reach while improving access and satisfaction. But success depends on more than tools. It requires ongoing learning, internal alignment, and attention to both legal and cultural expectations.

To move forward, review your current efforts. Identify what's missing. Then, prioritize changes that address measurable goals, like improved web visibility, higher appointment volume, or better patient reviews. Digital marketing is not a one-time project. It requires testing, refinement, and responsiveness to new technologies.

Stay current. Stay compliant. Stay connected. Doing so will help you grow your practice, protect your reputation, and better serve your patients.

C. Quinn, *The Digital Practice*, https://doi.org/10.1007/978-3-032-07745-5_22

Acronyms

2FA	Two-factor authentication. A form of MFA that uses two steps.
AI	Artificial intelligence. Software that performs tasks like pattern recognition, prediction, or generation.
API	Application programming interface. A set of rules that lets software systems exchange data.
AR	Augmented reality. Digital overlays on the real world viewed through a device.
CAC	Customer acquisition cost. Total spend divided by the number of new customers.
CCPA	California Consumer Privacy Act. State law on data rights for California residents.
CMS	Content management system. Software used to build and manage websites.
CPA	Cost per acquisition. Amount paid to generate one desired action, for example, a booked visit.
CPC	Cost per click. Average amount you pay for each ad click.
CPM	Cost per 1000 impressions. Price to show an ad 1000 times.
CRM	Customer relationship management. System for tracking contacts, activity, and pipeline.
CTOR	Click to open rate. Clicks divided by opens for an email.
CTR	Click-through rate. Clicks divided by impressions.
EHR	Electronic health record. Clinical system for patient data and workflows.
EMR	Electronic medical record. Single-practice clinical record system.
FAQ	Frequently asked questions. Structured Q and A used on websites and in schema.
FDA	Food and Drug Administration. U.S. agency that regulates medical devices and claims.
FTC	Federal Trade Commission. U.S. agency that enforces advertising and privacy rules.
GDPR	General Data Protection Regulation. EU privacy law that governs personal data use.

C. Quinn, *The Digital Practice*, https://doi.org/10.1007/978-3-032-07745-5

HIPAA	Health Insurance Portability and Accountability Act. U.S. law for protecting PHI.
HITECH	Health Information Technology for Economic and Clinical Health Act. U.S. law that strengthened HIPAA and breach rules.
IVR	Interactive voice response. Phone menus that route calls or collect information.
KPI	Key performance indicator. Metric tied to a business goal.
LLM	Large language model. AI model that generates or analyzes text.
MFA	Multi-factor authentication. Log in with two or more verification steps.
ML	Machine learning. Methods that learn patterns from data to make predictions.
MMS	Multimedia messaging service. Text messages with images or video.
NLP	Natural language processing. AI methods for analyzing or generating text.
NPS	Net Promoter Score. Measure of how likely users are to recommend your service.
OCR	Office for Civil Rights. HHS office that enforces HIPAA.
PAC	Patient acquisition cost. Total marketing spend divided by the number of new patients.
PHI	Protected health information. Identifiable health data covered by HIPAA.
PII	Personally identifiable information. Data that identifies a person, for example, name or email.
PPC	Pay per click. Online ads where you pay when a user clicks.
RAG	Retrieval augmented generation. Technique that lets an LLM use external sources for answers.
ROAS	Return on ad spend. Revenue from ads divided by ad spend.
ROI	Return on investment. Gain minus cost divided by cost.
SEM	Search engine marketing. Paid tactics to reach users on search engines.
SEO	Search engine optimization. Work to improve unpaid search visibility.
SGE	Search Generative Experience. Google's AI-generated responses within search results.
SMS	Short message service. Standard text messaging.
SSL	Secure sockets layer. Encryption for data in transit on the web. Often referenced as TLS.
SSO	Single sign on. One login grants access to multiple systems.
TLS	Transport layer security. Modern encryption protocol for data in transit.
UTM	Urchin Tracking Module. URL tags used to attribute traffic and conversions.
VR	Virtual reality. Immersive visual experience via headset or screen.

Glossary

Attribution model Method to assign credit for a conversion across touchpoints such as ad, email, or organic search.

Call to action (CTA) Clear instruction that tells the user what to do next, for example, Book an appointment.

Click-through rate Percentage of users who click after seeing an impression.

Click-to-open rate Percentage of users who click after opening an email.

Conversion Desired action, for example, request an appointment, call, or download.

Conversion funnel Path the user takes from awareness to action on your site or app.

Cookie consent User choice on tracking technologies recorded through a banner or preference center.

Data minimization Practice of collecting only the information needed for a stated purpose.

Deidentification Process to remove identifiers from data so it no longer links to a person.

FAQ schema Schema markup for common questions and answers on a page.

First-party data Information your organization collects directly from users with consent.

Geofencing Targeting users who enter a defined geographic boundary with ads or messages.

Geotargeting Ad targeting based on a user's location or a selected region.

Heatmap Visual summary of where users click or scroll on a page.

Knowledge panel Branded information box that appears on Google results for your organization.

Landing page Single page built to drive one action with focused content and a short form.

C. Quinn, *The Digital Practice*, https://doi.org/10.1007/978-3-032-07745-5

Lead magnet Resource offered in exchange for contact information, for example, a checklist.

Local Pack Three local results that appear on Google for location-based searches.

Multitouch attribution Attribution that credits more than one touchpoint for a conversion.

Open rate Percentage of emails opened out of all delivered emails.

Opt in User permission to receive messages from your organization.

Opt out User request to stop messages. It must take effect promptly.

Patient acquisition cost Total marketing spend divided by the number of new patients in the period.

Personalization Tailoring content or offers based on user data and behavior.

Retargeting Ads shown to users who visited your site or engaged with your content.

Return on ad spend Revenue from ads divided by ad spend for the same period.

Return on investment Gain minus cost divided by cost for a project or campaign.

Schema markup Structured data code that helps search engines understand your content.

Session replay Tool that shows how users moved through your site by recreating sessions.

Split test Another name for an A/B test that compares two versions of content.

Third-party data Information gathered by outside firms and sold or shared for targeting.

Tracking pixel Small code snippet that records visits or conversions for analytics.

UTM parameters Tags added to links that enable accurate campaign attribution in analytics.

Zero-party data Information a user proactively shares, such as preferences from a form.

Index

C. Quinn, *The Digital Practice*, https://doi.org/10.1007/978-3-032-07745-5

N

O

P

MIX
Papier aus verantwortungsvollen Quellen
Paper from responsible sources
FSC® C105338

If you have any concerns about our products,
you can contact us on
ProductSafety@springernature.com

In case Publisher is established outside the EU,
the EU authorized representative is:
Springer Nature Customer Service Center GmbH
Europaplatz 3, 69115 Heidelberg, Germany

Printed by Libri Plureos GmbH
in Hamburg, Germany